Rotational Moulding of Plastics
SECOND EDITION

The creation of the first edition of this book was supported by the Association of Rotational Molders, based in Chicago, Illinois, USA; and they are pleased to welcome the second edition. The use of their registered logo herein is by permission, and indicates support for and encouragement of its authorship, in which the Association had no direct participation nor right of control or approval.

POLYMER ENGINEERING SERIES

Series Editor: Professor R. J. Crawford
The Queen's University of Belfast, UK

2. **Rotational Moulding of Plastics,** SECOND EDITION
Edited by R. J. Crawford

Rotational Moulding of Plastics

SECOND EDITION

Edited by

R. J. Crawford

The Queen's University of Belfast, UK

RSP

RESEARCH STUDIES PRESS LTD.
Taunton, Somerset, England

JOHN WILEY & SONS INC.
New York • Chichester • Toronto • Brisbane • Singapore

RESEARCH STUDIES PRESS LTD.
24 Belvedere Road, Taunton, Somerset, England TA1 1HD

Reprinted August 1997

Marketing and Distribution:
Australia and New Zealand:
Jacaranda Wiley Ltd.
GPO Box 859, Brisbane, Queensland 4001, Australia

Canada:
JOHN WILEY & SONS CANADA LIMITED
22 Worcester Road, Rexdale, Ontario, Canada

Europe, Africa, Middle East and Japan:
JOHN WILEY & SONS LIMITED
Baffins Lane, Chichester, West Sussex, UK, PO19 1UD

North and South America:
JOHN WILEY & SONS INC.
605 Third Avenue, New York, NY 10158, USA

South East Asia:
JOHN WILEY & SONS (SEA) PTE LTD.
37 Jalan Pemimpin 05-04
Block B Union Industrial Building, Singapore 2057

Library of Congress Cataloging-in-Publication Data

Available

British Library Cataloguing in Publication Data
A catalogue record for this book is available from the British Library.

ISBN 0 86380 194 3 (Research Studies Press Ltd.) *[Identifies the book for orders except in America.]*
ISBN 0 471 96303 8 (John Wiley & Sons Inc.) *[Identifies the book for orders in USA.]*

Typeset by Bernie Nicholls Typesetting, Aylesbury, UK
Printed in Great Britain by SRP Ltd., Exeter

Foreword to the Second Edition

Rotational Molding is in a metamorphosis. Our industry is changing from a process where development and control were based largely on experience and trial and error, to a process based on the measurement and control of key parameters developed in a closed loop system of process control. The results have provided us with the ability to mold increasingly complex parts with significant improvements in appearance, tolerance control and lower factory costs. These lower factory costs are due to less scrap and better machine utilization.

Professor Roy Crawford and his group at The Queen's University of Belfast have been in the forefront of this evolution. The first edition of this book introduced many of these concepts and has become a primary reference tool for Rotational Molders throughout the world. This second edition adds to the study on process control and covers two new topics - Molding of Liquid Plastics and Mold Pressurization Technology.

The Association of Rotational Molders continues to support Professor Crawford's work. One of ARM's primary objectives is to encourage development of Rotational Molding technology and to keep our members apprised of the latest research. We look to future editions as Professor Crawford's work continues to identify new opportunities in Rotational Molding technology.

Bill Blaiklock
President
Association of Rotational Molders
March 1996

Foreword to the First Edition

Today, more is known about the many other plastics processes than is known about rotational molding. A wide spectrum of topics is covered in design manuals, handbooks, plastics trade shows and technical classes on processes other than rotational molding.

One of the goals of the rotational molding industry is to expand the advancement of this unique plastics process through its endeavors in the Association of Rotational Molders, ARM. ARM is an international trade association with members from over 40 countries.

The premise of ARM is to expand the industry, which has experienced rapid growth and acceptance in the past decade, through educational advancements, processing improvements, materials and technological breakthrough.

This book on rotational molding will expand your understanding of this unique process and will provide updated information to assist in the development of new products and applications for rotational molding. The authors are leaders in their specialty within the field of rotational molding.

Chuck Carlsen
Immediate Past President
Association of Rotational Molders
August 1992

List of Contributors

Beall, G. L.
Glen Beall Plastics, 32981 N. River Road, Libertyville, Illinois 60048, USA

Covington, W. H.
Ferry Industries, 1050 West Main Street, PO Box 249, Kent, Ohio 44240-0249, USA

Crawford, Professor R. J.
Department of Mechanical and Manufacturing Engineering, The Queen's University of Belfast, Ashby Institute, Stranmillis Road, Belfast, Northern Ireland BT9 5AH

Harkin-Jones, Dr. E.
Department of Chemical Engineering, The Queen's University of Belfast, Keir Building, Stranmillis Road, Belfast, Northern Ireland BT9 5AH

Kliene, R. I.
Chairman, Rotec Chemicals Ltd., 24 Norris Way, Wellingborough Road, Rushden, Northants., England NN10 6BP

Nugent, Dr. P. J.
now at REMCON Plastics Inc., 208 Chestnut Street, Reading, PA 19602-1809, USA

Petruccelli, F.
8 Wildwood Trail, Boonton, NJ 07005, USA

Scaccia, S. F.
Kelch Corporation, Aluminum Molds Division, W66 N622 Madison Avenue, Cedarburg, Wisconsin 53012, USA

Spence, Dr. A.
Department of Mechanical and Manufacturing Engineering, The Queen's University of Belfast, Ashby Institute, Stranmillis Road, Belfast, Northern Ireland BT9 5AH

Taylor, T. J.
T J Technical Services, 73 Midland Road, Raunds, Northants., England NN9 6JF

Wassell, G.
"Chellow-Dene", Church Lane, Flyford Flavell, Worcestershire, England WR7 4BZ

Contents

CHAPTER 1
Introduction to Rotational Moulding

R.J. Crawford

1. INTRODUCTION

Rotational moulding is a method for producing hollow plastic articles. The process was developed in the 1940's but in the early years it attracted little attention because it was regarded as a slow process which was restricted to a small number of plastics. However, over the past two decades, improvements in process control and developments with plastic powders have resulted in a very significant increase in usage of the process. Nowadays the advantages which it has to offer in terms of the economic production of quite complex, stress-free articles has made it a very competitive alternative to blow moulding and injection moulding.

Rotational moulding, also known as rotocasting or rotomoulding, is unique amongst plastics moulding processes because the heating, shaping and cooling of the plastic all take place inside the mould with no application of pressure. The concept is simple. A predetermined charge of cold plastic powder is placed in one 'half' of a cold mould - usually sheet steel or cast aluminium. The mould is then closed and rotated biaxially in a heated oven as shown in Fig. 1. As the metal mould becomes very hot, the plastic powder tumbling inside the mould starts to melt and coat the inside surface of the mould. When all the powder has melted, the mould is then transferred to a cooled environment. The biaxial rotation continues until the plastic has solidified. At this point the mould is opened and the product is removed. The article being produced need not be hollow since finishing operations such as cutting or sawing can be used to make, for example, right and left handed articles (1-6).

The earliest commercial applications of the concept of rotational moulding used vinyl plastisols to put a coating on the inside surface of a hollow metal mould. The use of these thick, viscous liquid dispersions led to the term 'slush moulding' being used for the process. Typical products were beach balls, toy dolls and artificial flowers. Today this type of material is still used to manufacture PVC film coatings for dashboards(7).

In the 1940's and early 1950's the process was slow to develop because the only other plastics suitable for use were some types of cellulosics. However, in the late 1950's the situation changed quite dramatically with the introduction of powdered

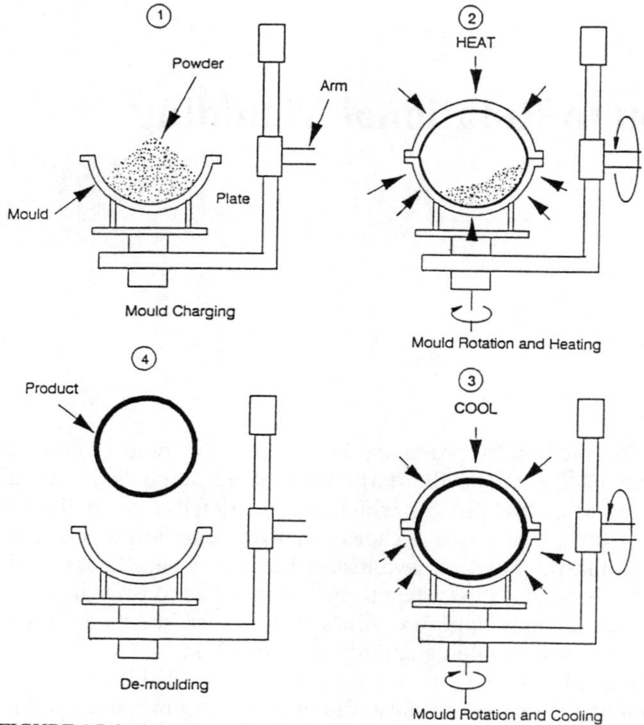

FIGURE 1 Principle of rotational moulding

polyethylene grades specifically developed for the process. Whole new market areas were opened up, although during the 1960's the toy industry was the largest industrial sector utilising the process. This was mainly because of the inherent low strength and stiffness of PVC and polyethylene. To improve this situation, cross-linkable grades of polyethylene as well as special rotomouldable grades of nylon, ABS, polycarbonate, HIPS, acetal, etc have been developed. However, even today polyethylene accounts for over 90% of the total tonnage of plastics used in roto-moulding. This is mainly because the powder flow, melt flow and thermal stability of this material are ideally suited to the process. Any shortcomings in mechanical properties such as stiffness can be readily overcome through the use of geometrical features, some of which are unique to rotomoulding(8).

An ever increasing awareness of the rotational moulding process has led designers to choose this process for some quite complex parts. As illustrated in Fig. 2, the growth in the usage of rotational moulded plastics is rapid, with typical rates being quoted as 10-12% per annum (1). Relative to blow moulding, rotational moulding permits much more complex, asymmetrical shapes. Wall thickness distribution is more uniform with rotomoulding and part size can go well beyond anything which is possible in blow moulding due to plasticisation limitations in the latter process.

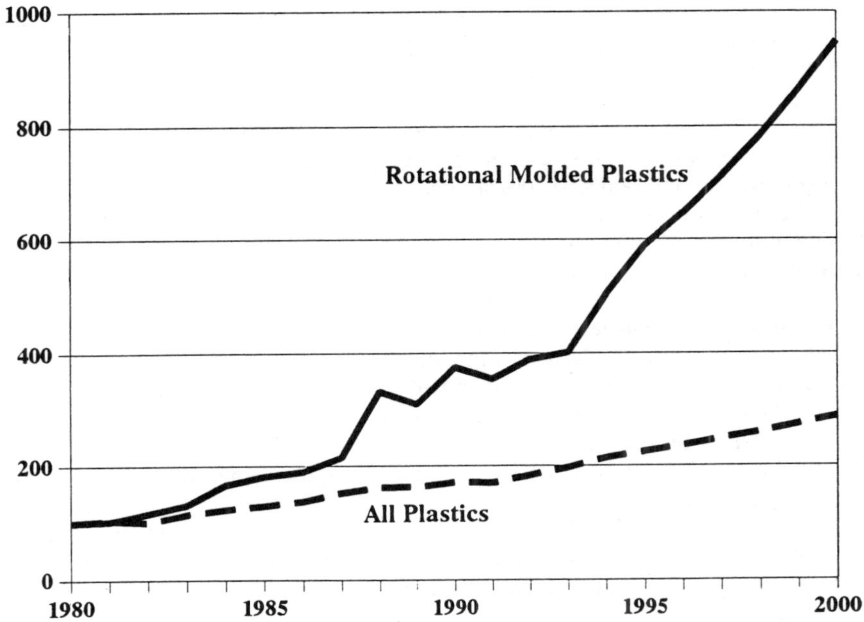

FIGURE 2 The growth of North American rotational moulding,1980-2000 (1980=100)

Relative to injection moulding, rotational moulding has the major advantages of very low mould cost and the ability to produce complex hollow articles. The cost difference between the moulds for the two processes becomes much more significant as part size increases. Hence, with mould costs typically one-fifth of those used in injection moulding, rotomoulding can easily overcome its slower cycle times through the use of multiple moulds.

The main advantages and disadvantages of rotational moulding are as follows:

1.1 Advantages of Rotational Moulding

The advantages and key features of rotational moulding may be summarised as follows:

1. Moulds are simple and relatively inexpensive. This arises from the fact that rotational moulding is a low pressure process and so high strength moulds are not required. Lead times for moulds can be as little as 2-4 weeks.
2. Moulds of quite different size and shape may be run simultaneously on the same equipment.
3. Articles with undercuts and intricate contours can be moulded easily. The process is particularly suited to large products, and special features such as double skinned mouldings can be readily moulded.

4. Parts have good wall thickness uniformity compared to those produced by processes such as blow-moulding or thermoforming.
5. Products can be virtually stress-free, provided precautions are taken. No weld lines or sprue marks are caused by processing.

A major feature of rotational moulding is undoubtedly its flexibility. A very wide variety of shapes and sizes can be moulded - ranging from tiny ear syringes to 100,000 litre storage tanks(6). Inserts may be moulded-in readily and surface textures such as wood grain, leather grain, etc may be accurately reproduced. Rotomoulding is ideally suited for economic short production runs and indeed it has an ever increasing role in prototype work for other processes.

1.2 Disadvantages of Rotational Moulding

1. Material costs are relatively high since most plastics are available as pellets and have to be reduced to a fine powder. The grinding stage adds significantly to the cost of the material, particularly as the quality of the grind is crucial to the performance of the material during moulding. In some cases the moulder will have his own grinding facility to improve the economics in regard to raw material cost.
2. The process is not suited for large production runs of small parts. This is because the plastic and the mould must be heated from room temperature to a relatively high temperature and then back to room temperature. This results in long cycle times.
3. Materials suited to rotational mouldings are limited. At present, polyethylene accounts for over 90% of the materials used but current research is extending the range of plastics which can be rotomoulded.
4. Loading and unloading is very labour intensive especially for complicated parts.
5. Bosses or solid ribs cannot be easily moulded. Hence, designers need to make clever use of other geometrical features in order to provide stiffness in mouldings.

2. MATERIALS

There are a number of fundamental characteristics which are desirable in a plastic which is to be rotationally moulded. Firstly, the flow characteristics of the powder are important to the way the plastic tumbles and distributes itself in the mould. Secondly, the flow characteristics of the melt are important if a uniform wall thickness is to be achieved and if the melt is to flow into complex shapes. In addition, the plastic must have a degree of thermal stability since it is to be subjected to elevated temperatures for prolonged periods of time. These characteristics can, to varying extents, be achieved in most thermoplastics. However, in many cases it is the economic considerations which rule out certain materials. In the following sections the plastics which can be rotationally moulded are described.

Polyolefins

For many years, polyethylene has had no competition as the most suitable material for rotational moulding. Its characteristics are dealt with in detail in Chapter 2 but the following section summarises some of the main points. Polyethylene is usually subdivided into one of three categories depending on density.

Material	Density (kg/m³)	% Crystallinity
Low density polyethylene	915-929	45-65
Medium density polyethylene	930-939	65-75
High density polyethylene	940-965	75-90

In general, properties such as stiffness, shrinkage, chemical resistance and hardness increase as the density increases. However, other properties such as impact strength and stress crack resistance will tend to decrease as density increases. These effects are illustrated in Fig. 3.

The fundamental differences between low and high density grades of polyethylene occur as a result of structural variations. In the high density grades, the molecular chains are essentially linear with few side branches so that the chains can pack closely together to provide more material per unit volume, ie higher density. In the low density grades, there are many side branches so that the chains cannot pack together closely. The fact that the chains in low density polyethylene (LDPE) are further apart and hence the forces between them are less strong gives the material its characteristic properties, ie lower strength, lower stiffness (flexibility) and ability to absorb impact (toughness.) In the past, if a stronger, stiffer material was required for rotomoulded products then it was necessary to move to a higher density grade but this meant a sacrifice of properties such as toughness and stress crack resistance.

In recent years, however, the development of linear low density grades of polyethylene (LLDPE) has made life much simpler for the designer and moulder. This new material has side branches which keep its density low but the branches

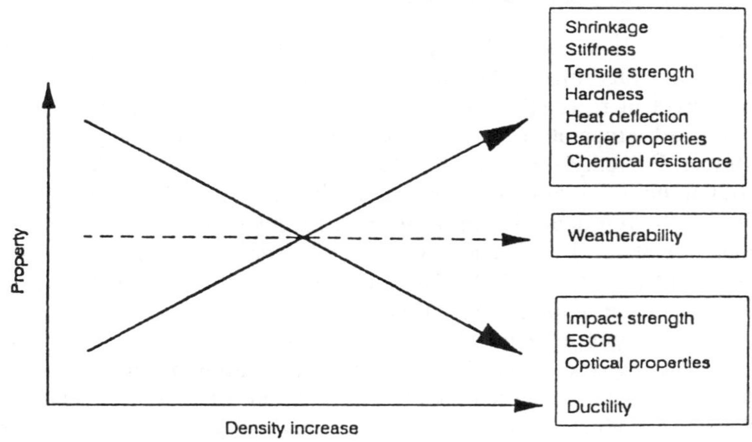

FIGURE 3 Effects of density changes on properties

are short and their occurrence can be well controlled. The density of LLDPE can be adjusted in the range 910-960 kg/m^3 The much more regular and refined structure of LLDPE means that it is stronger and stiffer than LDPE at the same density whilst retaining toughness and stress crack resistance.

In general, polyethylenes are resistant to most solvents at room temperature although aromatic and chlorinated hydrocarbons will cause swelling. LLDPE and HDPE have no known solvents at room temperature. LDPE is relatively unaffected by polar solvents, for example, alcohols, phenols, esters and ketones. However, caution is needed in regard to environmental stress cracking (ESC). This relates to the cracking which can occur in a material when it is under stress in the presence of a polar liquid or its vapours. Environmental stress cracking is also associated with detergents, silicone fluids, chloroform, xylene and paraffin. In polyethylenes the resistance to ESC decreases as the density increases and the high molecular weight grades give the best performance.

Polyethylenes are generally resistant to water, vegetable oils, alkalis and most concentrated acids at room temperature. However, their light (or ultra-violet) resistance is poor. The cheapest way to improve this is by the incorporation of carbon black but other methods are available. Water absorption is very low but will increase if carbon black is used.

LDPE is more permeable to gases and vapours than LLDPE and HDPE. The permeability for organic vapours is least for alcohols and then increases from acids to aldehydes and ketones, esters, ethers, hydrocarbons and halogenated hydrocarbons. Fluorine, a halogen, slowly attacks polyethylene and has been used in the past to improve its permeability. Polyethylene is attacked by fuming nitric acid and fuming sulphuric acid. It is also attacked slowly by chlorinating agents such as chlorosulphonic acid and phosgene.

The shrinkage of polyethylene is typically in the region of 2-5%. The higher values are associated with the higher density grades and are influenced by fillers such as carbon black.

Polyethylene is an easy flow material. This is quantified by a melt flow index (MFI) value – also called melt index (MI) or melt flow rate (MFR). The number quoted is the weight of material (in grams) which would flow through a standard orifice at a standard temperature in 10 minutes. Hence, an easy flow grade has a high MFI number. Polyethylenes can have MFI values in excess of 100, but commercial rotational moulding grades are usually in the range 2-8. If a lot of intricate mould detail is to be reproduced then a higher MFI grade is desirable. However, impact strength, stress crack resistance, weatherability and chemical resistance all decrease as the MFI increases (10,11). These effects are illustrated in Fig. 4. Low MFI values usually result in longer cycle times.

The molecular structure of polyethylene also has an influence on its properties and processing characteristics. The MFI value gives a rough indication of molecular weight or chain length. A material with a high melt index has shorter chains and a lower molecular weight or smaller molecules. This corresponds to the general deterioration in physical properties which was referred to above, as the MFI

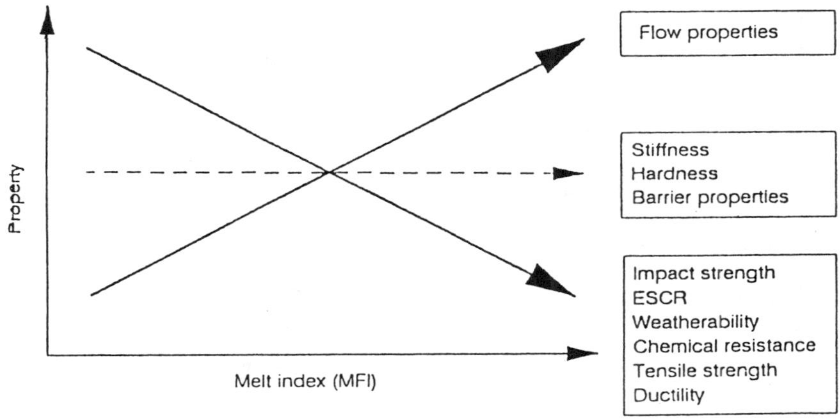

FIGURE 4 Effects of melt flow index on properties

increases (11). Unfortunately, however, materials with high molecular weights cannot be rotationally moulded because their melt characteristics are basically unsuitable for the fusion processes which occur. It is also important to note, however, that two plastics can have the same MFI but different molecular weight distribution (MWD). This characteristic of the plastic is dependent on the way in which the molecules are distributed. If the MWD is narrow then the mechanical properties (in particular, impact strength) will be better and warpage is reduced (11). On the other hand, a broad MWD results in better processability and improved environmental stress crack resistance (10).

As well as the rheological and chemical nature of the polyethylene, the physical nature of the powder can have an important effect on the performance during moulding and the properties of the final product. This is discussed in detail in Chapter 2. Normally the size of the powder particles varies from the finest (less than 150 microns) to about 500 microns.

The tumbling of the powder inside the mould results in a natural sorting of particle sizes. As a result, the particle size distribution is very important. Fine particles will tend to filter through the coarse particles and hence they will be the first to melt and fuse rapidly at the hot metal mould. This has the desirable effect of producing a fine texture surface with a minimum of voids. However, an excess of fine particles will cause 'bridging' - ie the small particles can melt and fuse together prematurely and prevent other particles from entering particular sections of the mould, for example, ribs or corners(9). In contrast, if there are too many large particles then melting will be slower and lumpiness will be evident at the inner surface of the product. The general effects of particle size are shown in Fig. 5.

It is evident therefore that the grinding of plastic powders from pellets is a very precise science. The principle of grinding is similar in most machines in that the pellets pass between stationary and rotating cutting plates and the particles are

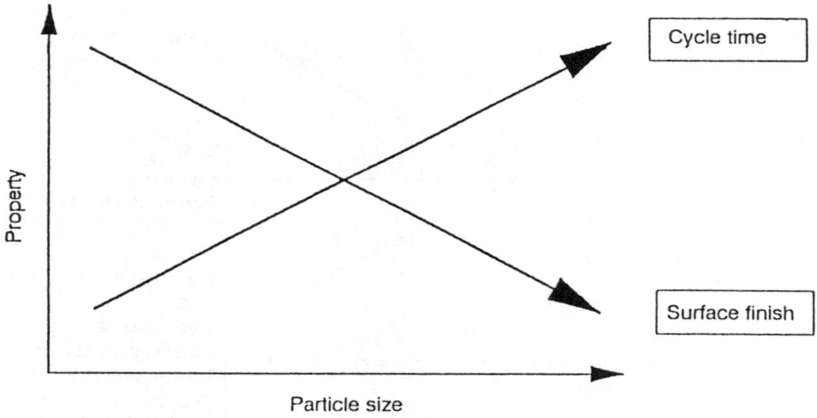

FIGURE 5 Effects of particle size on properties

swept out into an airstream at the periphery of the plates. The powder particles then pass through a series of sieves to control the particle size distribution. A crucial aspect of the grinding process is the shape of the resultant particle. Smooth particles are essential because if 'tails' are produced then (a) the 'bridging' effect referred to above will tend to occur, (b) there will be a high level of air bubbles trapped between the particles and (c) the particles will have a tendency to agglomerate into fluffy balls which melt late in the cycle and lead to rough, lumpy areas in the moulding. Particles with 'tails' will also result in a low bulk density - that is, a smaller weight of powder can be put into a fixed volume.

The polyethylene family of rotomouldable materials is large. As well as the low, linear low, medium and high density grades referred to already it is possible to have cross-linked grades, and alloys or hybrids such as ethylene vinyl acetate (EVA) or ethylene butyl acrylate (EBA) (12-14). The former, cross-linked, grades provide materials which are very strong and stiff and have excellent stress crack resistance. Three common ways in which the cross-linking can be achieved are:

1. the addition (into the polymer) of a heat activated radical source, most commonly an organic peroxide
2. incorporation of functionality by copolymerisation or by chemical grafting on to the polymer chains. One of the most commercially important examples of this technique with polyethylene is the free radical initiated grafting of vinyl silane molecules on to polyethylene. An advantage of this approach is that cross-linking can be activated by a secondary process on the product - involving the use of water rather than heat.
3. cross-linking by irradiation. As with (2) the cross-linking is achieved after moulding by, in this case, exposing the product to a high energy radiation source.

The use of cross-linked polyethylenes is becoming more popular in rotational moulding due to the enhanced properties of these materials. The improved environmental stress crack resistance arises because although the mechanisms of attack are the same, the cross-links inhibit slipping of the tie molecules. There is therefore improved resistance to detergents and solvents. The effect of oxidisers is much the same as for normal polyethylenes but the functional groups formed during chemical cross-linking provide more sites for potential attack.

The hybrid materials such as EVA and EBA are becoming more common when very flexible, tough products are required. These materials can be difficult to mould without pin-holes although new developments such as those described in Chapter 10 can overcome these problems.

In addition, commercial rotomouldable grades of polypropylene are now available and these offer the strength and stiffness of the very highest density grades of polyethylene but at a density in the region of 900-910 kg/m^3. Their stress crack and chemical resistance are also excellent but care is needed when good toughness is required.

Vinyl Plastics

Liquid plastisol resins are dispersions of polyvinyl chloride in plasticisers and they usually include other compounding ingredients such as stabilisers, fillers and pigments (15-18). Plastisols were used in Europe in the 1930's and became used as rotomoulding materials in the 1940's. For about 20 years they were the only materials suitable for this process but in the 1960's they became superseded by polyethylene. However, plastisols still have an important part to play in this industry. Nowadays they are used for a wide range of flexible products, particularly toys. They are also very popular in car interiors as the soft flexible covering for dashboards and door panels. This soft skin is produced by a technique known as Slush Moulding. The liquid plastisol is used to coat the inside of a biaxially rotated heated mould. When the desired skin thickness has been achieved, the excess plastisol is poured from the mould and the skin moulding is peeled off. It is then placed over a foamed plastic or other backing material to provide support (7).

In the early 1940's the possibility of using powdered polyvinyl chloride plastics as rotomoulding materials became apparent. They offer the normal advantages of PVC in terms of wide range of stiffness possibilities, very good chemical resistance and above average weatherability. Nowadays the areas of applications for rotomoulded PVC products are increasing although they tend to be mainly in the toys and recreational sectors (17,18).

To rotomould PVC plastisols or powders successfully, it is essential to have three ingredients. These are the basic PVC resin, a plasticiser and a heat stabiliser. Typically these are combined at levels of 100 parts by weight of PVC, 25 to 100 parts of plasticiser and 1 to 5 parts of stabiliser. The basic PVC resin cannot be rotomoulded on its own because it is very susceptible to thermal degradation. The plasticiser is used to soften the plastic and impart flexibility to the product. The heat stabiliser is necessary because the molecular structure of PVC will start to break down at normal processing temperatures. This results in discolouration and a drastic loss of physical properties. The stabiliser has the effect of reacting with the molecular structure and prevents the propagation of double bonds. Typical stabilisers are mixtures of fatty acid salts of barium, cadmium, zinc or calcium but other compounds are also very effective (18).

The formulation of PVC plastisols is extremely important because it has an influence on the processing behaviour and the properties of the final product. In particular, the choice of plasticiser is crucial to a successful product. Plasticisers can be classified according to how compatible they are with the PVC resin. If the plasticiser is not compatible then it may perform a short-term role but will migrate out of the PVC. If the plasticiser is compatible over a wide range of temperatures then it is considered to be a primary plasticiser, such as those based on esters. PVC is also popular in food packaging applications and in these cases it is necessary to have special approval from the food authorities - in particular to satisfy stringent requirements in regard to zero or low migration of the plastic or its additive into the food. In these applications special types of plasticisers must be used (18). Even for non-food applications the plasticiser needs to be carefully selected because, for example, some types can make the PVC more flammable. As a result, the moulder is advised to liaise closely with the resin supplier to select the appropriate grade for a specific application.

Other Plastics

A wide range of other thermoplastics can be rotationally moulded. These are conveniently divided into the categories of those which are used on a commercial basis and those which are not in regular use but which offer significant potential. Materials which are used commercially are polyester, polycarbonate, nylon 6, nylon 11, nylon 12, polyvinylidene fluoride and fluoropolymers such as ECTFE (4,12,19,23). The growing importance of nylon as a rotomoulding material means that it merits special treatment in Chapter 3. The thermoplastic polyester materials have superb toughness and long term resistance to water. They have been used successfully for tank liners, water beds and flexible containers. Polycarbonate has excellent stiffness, toughness and resistance to creep. It can be moulded clear and is popular for light globes, engine covers for snowmobiles and storage containers. The Nylon range of materials also offer very good strength, stiffness and toughness (relative to polyethylene). They have excellent chemical resistance and are the most suitable rotomoulding material for fuel tanks. They normally require an inert atmosphere in the mould during processing to prevent oxidation/degradation effects. The fluorocarbon family of plastics offer superb chemical resistance. Ethylene-Chlorotrifluoroethylene (C-CTFE) is the material which is most commonly used for rotomoulding.

Other thermoplastics which can be rotomoulded, although they are not in common use, include cellulosics, ABS, polyurethane, acetal, polyphenylene oxide, polyphenylene sulphide and ionomers (4,23-25). These materials, mostly regarded as engineering plastics, permit rotomoulding to move into new market areas, eg electrical housings, business machines and light engineering products. The typical characteristics of the engineering plastics are that they are strong and stiff, and good dimensional stability can be achieved (in the amorphous resins). However, to counter these attractive features they usually only have moderate chemical resistance, are difficult to rotationally mould and are relatively expensive.

Reactive Materials

It is important to note that a number of reactive materials can be rotationally moulded. These include thermosetting materials such as polyester, silicone rubber, reactive nylons and polyurethanes. Foaming grades of thermoplastics could also be put in this category (26-29). There are a number of important differences between the behaviour of reactive resins and that of the powdered thermoplastic resins more commonly used in the rotational moulding process. For example, most of the reactive resins are placed in the mould as low viscosity liquids. In addition, since the conversion from liquid to solid occurs as a result of a chemical reaction, there is no need for the heating and cooling cycle which is normally necessary. In the case of thermosetting polyesters, the mould is biaxially rotated at room temperature. In the case of reactive nylons, the process is most effectively operated at constant mould temperatures, in the region of 100°C. It must be said that the rotational moulding of reactive liquids is not yet very common due primarily to the practical difficulties of working with exothermic liquids with time dependent viscosities. However, these systems do have many significant advantages to offer in terms of stiff/strong moulding with intricate detail. It is likely therefore that their usage will increase very considerably in the future. Due to the low viscosity of most reactive resins, there is also the exciting possibility of using them to impregnate fibre preforms placed in the mould. The moulding of liquid plastics is considered further in Chapter 11.

General Information

Further information on the range of materials available to rotomoulders may be found in the ARM rotomoulding library(30). All rotomoulding materials require additives to enhance their performance both at the moulding stage and as end products. Pigments are usually required to colour the plastic. Melt compounding of the pigment before the grinding stage to produce the powder, gives the best results but it is more expensive. As a result, dry blending of the pigment with the plastic powder prior to charging the mould is quite commonly used to reduce costs. However, this approach has a number of disadvantages in that it results in lower impact strength, and environmental stress crack resistance is reduced in the moulding.

Carbon black, at levels of about 1½%-2½%, is often used to reduce photocatalytic oxidation, ie embrittlement due to exposure to sunlight. UV stabilisers are also essential for products which will be used outdoors for prolonged periods. The most efficient systems for polyethylenes are known as HALS (hindered amine light stabilisers). Dry blending of these is not effective and the best results can only be achieved by incorporation at the reactor stage(9). The protection offered by UV stabilisers is thickness dependent, and thin mouldings will embrittle more quickly than thick mouldings. Anti-static agents are also used in rotomoulding. Effectively they improve the electrical conductivity of the surface of the moulding so that static pick-up is reduced. This means that there will be less dirt and dust attracted to the surface of the mouldings. Anti-static agents are basically 'soapy' chemicals which are incompatible with the plastic. Hence they tend to migrate to the surface and their high moisture content improves the electrical conductivity of the surface layer.

Environmental Stress Cracking (ESC)
The resistance of plastics to attack by liquids has always been of keen interest to rotational moulders. Environmental Stress Crack Resistance (ESCR) is defined in ASTM D-1693 and is referred to in Chapter 2 in relation to polyethylenes (66). However, other materials are also susceptible and so due care is always needed when a plastic is being considered for a new application area. It should be noted that for ESC to occur it is necessary to have stress in the plastic as well as the presence of a stress cracking agent. The following guide has been prepared by Exxon Chemicals to assist in choosing the correct plastic for a particular application.

Performance	**Environment**		
	Surfactants	**Fuels**	**Oxidizers**
Best Least Resistant	. Polypropylene . Crosslink PE . PE . Polycarbonate	. Nylon 6 . Nylon 11, 12 . Rigid PVC . Crosslink PE . Fluorinated PE . HDPE . LLDPE . Polycarbonate	. PE . Polypropylene . Crosslink PE . Polycarbonate
General Comments	. Generally Higher MW Helps	. Generally Higher Crystallinity and Higher MW Helps	. Generally Lower MI or Higher MW Helps . Inertness (Non-Polar Surface) Key Feature

3. ROTATIONAL MOULDING MACHINES

The basic requirements of a rotational moulding machine are that the mould is first heated and then cooled whilst it is rotating, usually in a biaxial manner. However, there are many different types of machine configuration and a variety of ways of supplying the heat and cooling the mould.

One of the earliest types of machine was referred to as the 'Rock and Roll' machine. This name related to the motion of the mould in that it rotated about a spindle which rocks back and forward. Hence, this machine does not provide a full double axis motion since the mould only rotates through 360° on one axis (the spindle). The angle of the rocking motion of the spindle can be altered depending on the shape of the product. The mould is heated directly by a series of gas jets placed below the mould. In the cooling stage, the gas flames are extinguished and the mould continues to rotate as it is cooled by air and/or water spray. This type of machine is restricted in the shapes which it can produce and it was soon recognised that full biaxial rotation is needed for more complex product shapes.

The way in which biaxial rotation of the mould is achieved is shown in Fig. 6. This illustrates the straight arm and cranked arm types of mould mounting. Although, in the past, attempts were made to heat the mould with hot oil or hot salt in a jacket surrounding the mould, these methods are not common nowadays. They offer very efficient heat transfer and so are ideal for high temperature materials such as polycarbonate but they do introduce additional maintenance problems. In modern rotational moulding machines, an oven is used to heat the mould. In most cases the oven is air heated and uses gas, oil or electricity as the heat source. Electricity is clean but slow, oil can be messy and requires regular maintenance whereas gas has proved to be the most popular because it is convenient and efficient.

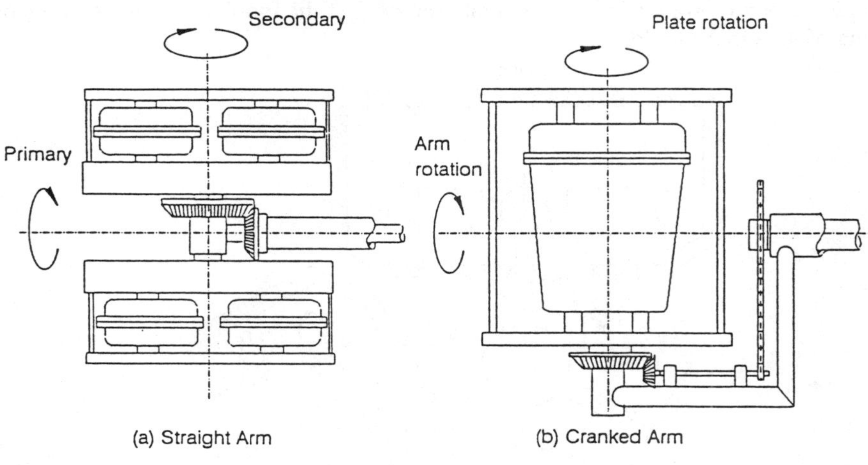

FIGURE 6 Two types of mould mounting

There are many options for transferring the mould from the charging area to the heating area (oven) and then to the cooling area. The three main categories of modern machine are:

(a) shuttle type machines
(b) fixed arm carousel machines
(c) independent arm carousel machines.

The simplest concept is a shuttle type machine where the mould runs on a track between the three stations (Fig. 7). For fast production rates, however, it is more common to have a multiple arm carousel machine.

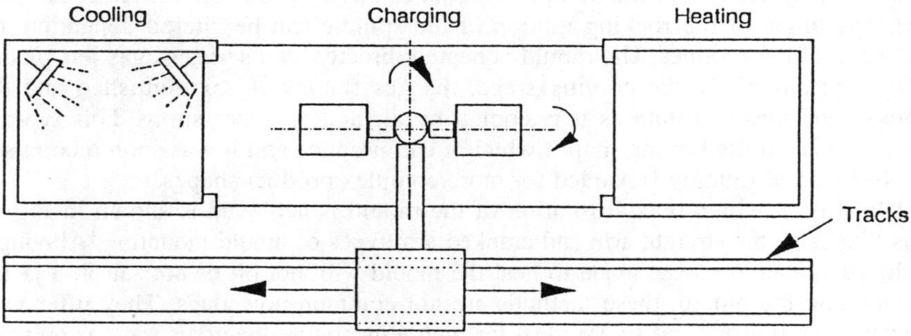

FIGURE 7 Shuttle type rotational moulding machine

The simplest form of this is a 3 arm machine as illustrated in Fig. 8. In this case, one set of moulds is being charged with powder and one set of moulds is in the oven while a third set of moulds is in the cooling bay. At pre-set times the arms index around through 120° to facilitate removal of the mouldings and heating/cooling of the other moulds.

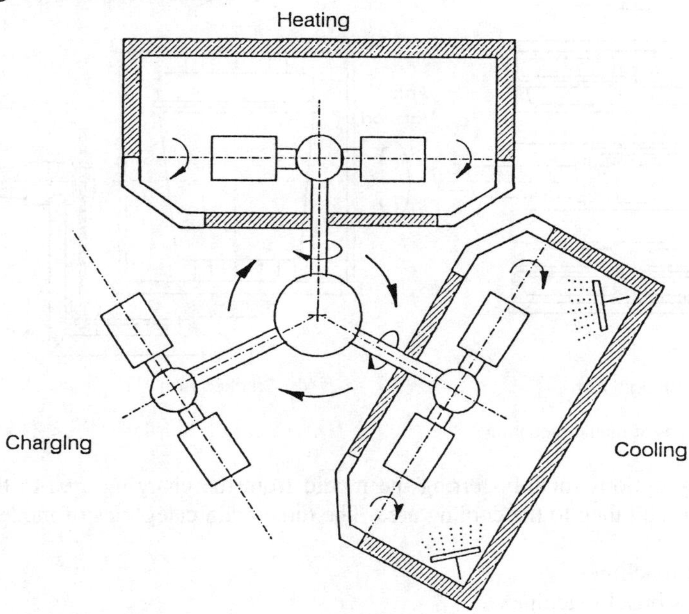

FIGURE 8 Carousel type rotational moulding machine

In recent years this type of machine has been made much more flexible in that the arms are not rigidly fixed together. This means that programmable logic controllers can be used to index each arm individually at the optimum time. Hence the movement of the arms is not controlled by the slowest event, as happens on the fixed arm machines. Further refinements of the independent arm machines include the introduction of additional ovens and cooling areas so that production rates can be pushed to levels which are competitive with blow moulding.

One other type of machine which is worth mentioning because of its increasing popularity is the Clam Shell machine. This is different to the machines described above in that the charging, heating and cooling all take place in the same chamber. This chamber opens like a clam shell so that the mould can be charged with powder. The 'shell' then closes around the mould which rotates biaxially in a heated environment (23,31). At a pre-set time the cooling cycle is activated and eventually the clam shell opens again to permit the moulding to be removed.

4. MOULDS

The moulds used in rotational moulding are shell-like. They define the outside shape of the product but do not have an internal core. There is a wide choice of materials available for moulds, and the construction method may involve fabrication, electroforming or casting (17,23,32-34).

Cast aluminium is used most extensively for multi-cavity moulds and for complex shapes. Typical wall thicknesses vary from 5-10 mm. The cost of one cavity is high due to the need for a pattern but subsequent cavities cost progressively less. Electroformed moulds are favoured for vinyl plastisol products. The reproduction of detail is very precise by this method. For very flexible products (eg. dolls' heads) there is no need for a parting line. Fabricated sheet metal moulds are generally the most economical and are almost always used for very large products.

The sheet metals used for mould fabrication can be aluminium, mild steel or stainless steel. Machined moulds are used occasionally but their costs tend to be prohibitive. Chapters 5 and 6 deal with the design aspects of sheet metal and cast aluminium moulds.

The basic requirements of a mould are as follows:

(a) It should have a good thermal conductivity so that the heat is transferred to (and from) the plastic as quickly as possible.

(b) The mould should be able to withstand, without warping, the thermal cycling of the heating and cooling stages of the process.

(c) There should be the provision of quick release clamps to keep the halves of the mould tightly closed during the heating and cooling stages but which facilitate fast opening of the mould at the end of the cycle.

(d) The mounting of the mould on the arm/plate should not prevent the free passage of air over the entire surface of the mould - otherwise hot spots will occur and cause an uneven distribution of plastic.

In addition, most moulds have a vent pipe. The function of this is often misunderstood and problems which are attributed to mould venting are often more directly related to other factors. For example, if the mould has a vent but this becomes blocked at some stage during the heating cycle, then there can be problems with warpage/distortion of the moulded product. This is because, during cooling, there is insufficient air to fill the space inside the mould at the lower temperature. This causes a partial vacuum inside the moulding which can be sufficient to pull the product away from the mould. If there had been no venting throughout the heating and cooling cycle then the reduced internal pressure would not have occurred and hence there would be no distortion - assuming that the mating surfaces of the mould 'halves' are airtight.

It has to be recognised, of course, that in the majority of cases the mating surfaces of the mould are not airtight. This can cause two problems. Firstly, air can escape during the heating stage and result in a partial vacuum effect during cooling as described above. Secondly, if there is no vent and the parting line is not airtight, it is possible that the molten plastic can be squeezed out to cause 'flash' and possible defects in the product wall at the parting line.

Hence, to overcome these problems it is normal to have an adequate vent (10-14 mm per m^3 of mould volume) to ensure equalisation of internal and external pressure in the mould.

When selecting a manufacturing method for the mould, the six key factors are Shape, Size, Quantity, Appearance, Heat Transfer and Weight Capacity of machine. The effects of Shape (ie complexity) and Size have been referred to above. Complex shapes usually have to be cast whereas large products usually necessitate a sheet metal mould. The Quantity required is important because, for example, multiple cavities may be required to meet the production rate. If this is the case then casting is normally the best choice. The Appearance of the product, in terms of aesthetics, dimensional stability etc is also an important consideration. Special textures and surface detail can be difficult to produce in sheet metal moulds. On the other hand, cast aluminium moulds tend to be softer and more susceptible to surface damage. Electroformed nickel moulds, in particular, have limited longevity.

The Heat Transfer characteristics of a mould can be important if it is to be mounted with other products on a carousel type machine. It is generally desirable to balance the cycle times of all the products in order to avoid over- or under-heating. Finally, the Weight Capacity of the moulding machine is an important practical consideration and care must be taken that the mould material will not cause this to be exceeded(34).

Mould finishing depends to a very large extent on the nature of the product. For many types of industrial mouldings, a minimum of mould finishing is required. Machine marks may be left in the mould and it may not be necessary for welds to be totally ground out. For more demanding customer applications, the mould may have to be polished to a fine emery or a high gloss finish to remove all traces of metal working, welding, etc. In other cases, a textured finish may be called for. This will have to be cast in if the texture is of a special type, or it could be introduced by etching or shot (sand) blasting.

In general, moulds for rotomoulded products are not required to have deep cavities or ribs because the plastic powder would not be able to penetrate these sections. However, special features such as 'kiss-off' points are quite commonly used to provide stiffening of polyethylene products (see Chapter 7). Within the past 10 years the inclusion of metal inserts in rotomoulded products has also been perfected. This is now quite common in tanks, for example, to facilitate brass pipe connections. In some cases, special spring loaded jigs are used to keep the insert in place as the plastic cools and contracts(34).

An important aspect of the rotational moulding process is the release of the product from the mould. If precautions are not taken then it is possible for the product to stick to the mould through either mechanical or chemical adhesion(9). To overcome this problem it is normal to use release agents on the mould. The operation of these is described in detail in Chapter 2.

Generally there are three classes of release agent:

1. Transient Release Agents (eg. Silicone emulsions, oils or waxes)
2. Semi-Permanent Release Agents (eg. Cross-linked siloxanes, silizanes)
3. Permanent Release Agents (eg. Baked-on Fluoropolymer eg. Teflon)

The transient release agents have the advantages that they are low cost, easy to use and possess good release properties. However, they are migratory which can affect finishing (eg painting), they require frequent application and they are moderately affected by the high temperatures used in rotomoulding. The semi-permanent release agents are not quite so user-friendly but they are nonmigratory, are not affected by the high process temperatures and provide a good, consistent surface finish on the product. The permanent release agents are an attractive idea (like Teflon coated frying pans) but they require special expertise in their application, they can cause premature release of the moulded part (leads to warpage) and they do tend to transfer off the mould, albeit in small amounts.

5. GENERAL DESIGN PROCEDURES FOR PLASTICS

The detailed design of rotomoulded products is considered in later chapters. However, it is appropriate at this stage to look at the general mechanical design of plastic products because some of the procedures are different to those used in metals. It is generally accepted that in any design, the following factors need to be considered.

(a) Mechanical properties, ie, strength, stiffness, specific strength and stiffness, fatigue and toughness, and the influence of high or low temperatures on these properties;
(b) Corrosion susceptibility and degradation;
(c) Wear resistance and frictional properties;
(d) Special properties, for example, thermal, electrical, optical and magnetic properties, damping capacity, permeability, etc;
(e) Moulding and/or other methods of fabrication;
(f) Total costs attributable to the selected materials and manufacturing route.

Of these factors, there are some which require particular attention for plastics and these will be considered in the following sections.

5.1 Deformation Behaviour of Plastics

Short-term Behaviour

Thermoplastics are viscoelastic materials, which means that their mechanical properties reflect the characteristics of both viscous liquids and elastic solids. Thus when a thermoplastic is stressed it responds by exhibiting viscous flow (which dissipates energy) and by elastic deplacement (which stores energy). The properties of viscoelastic materials are time, temperature and strain-rate dependent. Nevertheless the conventional stress-strain test is frequently used to describe the short-term mechanical behaviour of plastics. It should always be remembered, however, that due to the time-dependent behaviour of plastics, short-term data should not be used in design calculations. Their function is to provide a means of carrying out an initial sorting of materials. Long-term design calculations must be carried out as described later using data obtained from long-term tests.

In many respects the stress-strain characteristic for a plastic is similar to that for a metal. From Fig. 9 it may be seen that at low strains there is a linear elastic region whereas at high strains there is a non-linear relationship between stress and strain and there is a permanent element to the strain. In the absence of any specific information for a particular plastic, design strains should normally be limited to 1%. Lower values are recommended for the thermoplastics such as acrylic, polystyrene and polycarbonate and values of 0.2-0.3% should be used for thermosets. Good quality polyethylene, polypropylene and acetal mouldings can be used at up to 2.5% strain.

However, whereas with metals one would expect the characteristic illustrated in Fig. 9 to be unique for a particular metal this is not so with plastics. For any selected

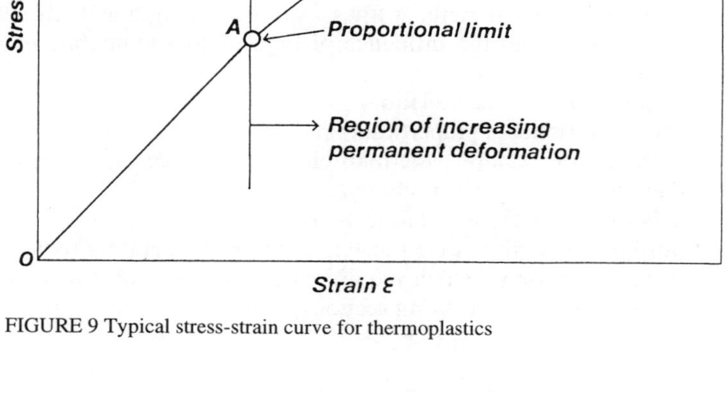

FIGURE 9 Typical stress-strain curve for thermoplastics

plastic the shape of this characteristic will be dependent on, in particular, the strain rate and the temperature. Higher strain rates cause a plastic to be stiffer and stronger but to possess less ductility. Moderate increases in ambient temperature will cause the plastic to become more flexible so that its modulus can decrease markedly. It is important to realise also that within the range of grades that exist for a particular plastic, there can be significant differences in mechanical properties. For example, with polypropylene, for each 1 kg/m^3 change in density there is a corresponding 4% change in modulus. And quite apart from density changes, a grade of material selected for some specific desirable feature (eg. high impact strength) will undoubtedly exhibit a decrease in some other property of the material (eg. tensile strength).

The stiffness of a plastic may be expressed in terms of a Modulus of Elasticity. Most values of elastic modulus quoted in technical literature represent the slope of a tangent to the stress-strain curve at the origin. This is often referred to as Young's modulus, E, but it should be remembered that for a plastic this will not be a constant and, as mentioned earlier, is only useful for quality control purposes, not for design. Since the tangent modulus at the origin is sometimes difficult to determine precisely, a Secant Modulus is often quoted to remove any ambiguity. A selected strain value of, say, 2% (point C', Fig. 10) enables a precise point, C, on the stress-strain curve to be identified. The slope of a line through C and O is the secant modulus. Typical short-term mechanical properties of plastics are readily available from the material suppliers or standard texts (61-63).

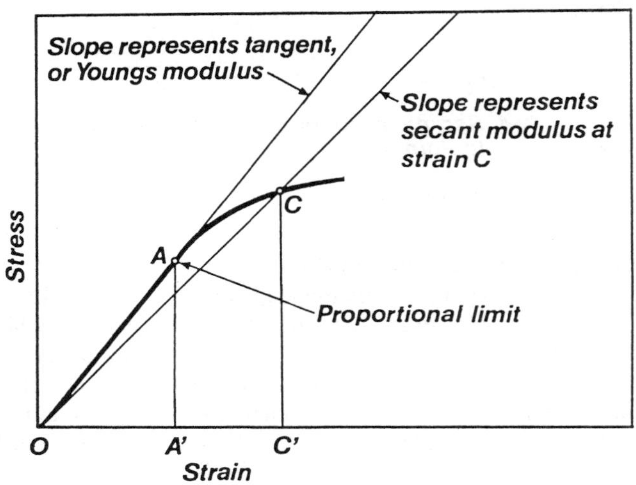

FIGURE 10 Secant and tangent moduli for plastics

Long-term Behaviour

Plastics exhibit a time-dependent strain response to a constant applied stress. This behaviour is called creep. In a similar fashion if the stress on a plastic is removed it exhibits a time-dependent recovery of strain back towards its original dimensions. This is illustrated in Fig. 11. Another important consequence of the viscoelastic nature of plastics is that if they are subjected to a particular strain and this strain is held constant it is found that as time progresses the stress necessary to maintain this strain decreases. This is termed Stress Relaxation and is of vital importance in the design of gaskets, seals, springs and snap-fit assemblies. The influence which these time-dependent phenomena have on design procedures for plastics is extremely important and this will be considered in the next section.

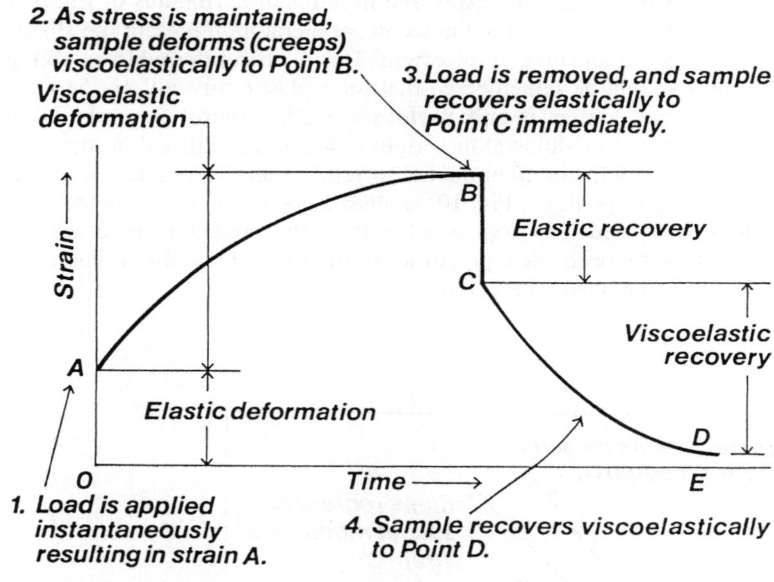

FIGURE 11 Typical creep and recovery behaviour of plastics

Design Methods for Plastics using Deformation Data

The most common method of displaying the interdependence of stress, strain and time is by means of creep curves. However, there are also other methods which may be more useful in particular applications (see BS 4618). The first of these is obtained by taking a constant strain section through the creep curves and re-plotting as stress versus time (Fig. 12(a)). This is termed an Isometric Curve and is an indication of the relaxation of stress in the material when the strain is kept constant. In addition, if the vertical axis (stress) is divided by the strain (0.015 in this case) then we obtain a graph of modulus against time (Fig. 12(b)). This is a good illustration of the time-dependent variation of modulus referred to earlier.

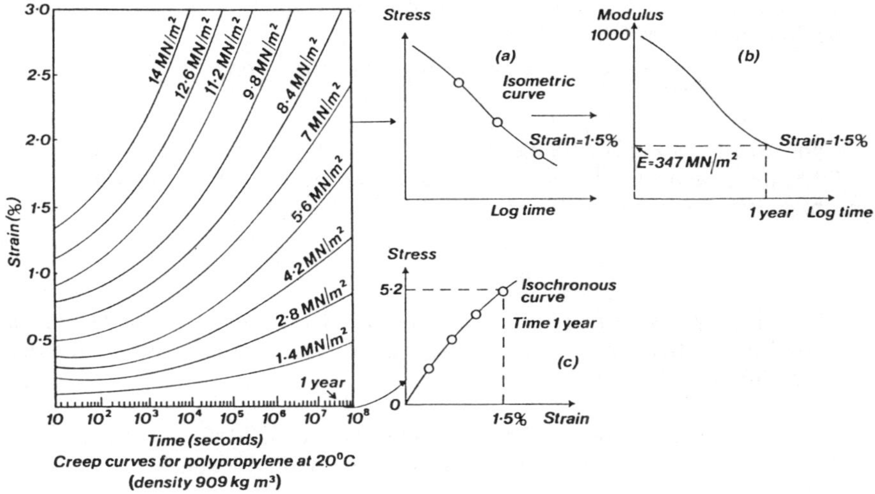

FIGURE 12 Construction of isometric and isochronous curves

A further graph may be obtained by taking a constant time section through the creep curves and plotting stress versus strain as shown in Fig. 12(c). This is termed an Isochronous Curve and it is particularly popular because, as it is a graph of stress against strain, it presents information in a form which is familiar to many designers. It may be used like any other stress/strain graph so long as one remembers that the data contained are time-dependent.

The way in which creep curves are used in design calculations is illustrated in the following example.

Example - A polypropylene beam 12 mm wide by 3 mm deep and 50 mm long is simply supported at each end and subjected to a point load at mid-span. If the beam is to be designed for a maximum strain of 1.5%, determine the largest load which may be applied at mid-span so that the deflection of the beam does not exceed 5 mm over a period of 1 year.

Solution - The second moment of area, I, for the beam is

$$I = \frac{\mathrm{bd}^3}{12} = \frac{12 \times 3^3}{12} = 27\ \mathrm{mm}^4$$

The central deflection for a beam loaded in this way is given by

$$\delta = \frac{WL^3}{48EI}$$

In order to find a value of W for the conditions specified it is necessary to have a value for the modulus E. This may be obtained from either the 1.5% modulus time graph (Fig. 12(b)) or the 1 year isochronous graph (Fig. 12(c)) for polypropylene. In either case E = 347 MN/m² so

$$W = \frac{48EI\delta}{L^3} = \frac{48 \times 347 \times 10^6 \times 27 \times 10^{-12} \times 0.005}{(0.05)^3} = 18 \text{ N}$$

Note that from Fig. 12(b), the short-term modulus of the polypropylene is about 1,000 MN/m² and so it drops off to about ⅓ of this value in 1 year. Thus when the beam in this example is first loaded, its deflection will be

$$\delta = \frac{WL^3}{48EI} = \frac{18\ (0.05)^3}{48 \times 1 \times 10^9 \times 27 \times 10^{-12}} = 1.7 \text{ mm}$$

As time progresses, the beam deflection will increase to reach the design condition of 5 mm deflection at 1 year. Table 1 gives typical values for the time-dependent moduli for a range of plastics.

Table 1 - Tensile Creep Modulus GN/m² at 20°C, 1% Strain
(Figures in parenthesis are extrapolated values)

Time	Polythene (LDPE)	Polythene (HD1PE)	Propylene Homopolymer Density 909 kgm³	Acetal Copolymer	Poly-methyl methacrylate	Rigid PVC	Nylon 66 (dry)*	Nylon 66+33% Glass-fibre (dry)**
1 Hour	0.143	0.89	0.80	1.93	2.34	2.76	2.91	9.2
1 Day	0.112	0.51	0.60	1.49	2.05	2.55	2.66	8.74
1 week	0.101	0.394	0.505	1.25	1.86	2.31	2.22	7.85
30 Days	0.098	(0.33)	0.444	1.11	1.71	2.07	1.88	7.10
1 Year	0.083	(0.214)	0.355	0.90	1.42	1.67	1.43	6.10
3 Years	(0.077)	(0.199)	0.321	0.82	0.29	1.45	(1.20)	(5.48)
10 Years	(0.075)	(0.181)	(0.287)	(0.75)	(0.16)	(1.35)	(1.11)	(5.26)

* *Modulus at ambient humidity is 67% of dry modulus; wet, 25% of dry modulus
*Modulus at ambient humidity is 67% of dry modulus; wet, 50% of dry modulus

Multi-layer mouldings

Rotational moulders are to a large extent obliged to use polyethylene on the grounds of availability and price. As this material has a relatively low inherent stiffness, designers must make very effective use of shape in order to get good rigidity in the moulded product. With injection moulded or extruded products, ribs and corrugations can be used to very good effect but such geometrical stiffeners are difficult by rotational moulding. Fortunately, similar results can also be achieved by using sandwich structures. Nowadays it is possible to produce rotationally moulded articles which have a solid skin and a foamed core. The foamed core may be introduced after moulding, eg. the injection of polyurethane foam is common if good thermal insulation properties are to be achieved. Alternatively, the foam may be of the same material as the solid skin and is introduced during the moulding process - either using a drop box or by using the newly developed one-shot foaming materials.

Thus the moulding of foam sandwich structures is relatively straightforward but the design process is complicated by the fact that the material properties are changing across the section of the moulding. However, this situation can be allowed for in a simple manner and the method is illustrated below. The procedure also works for situations where the sandwich section is of two different solid materials. For convenience the latter case will be illustrated first.

Design of Solid A/Solid B Sandwich Sections

Consider a cross-section of a rotationally molded product as shown in Fig. 13(a). This could be a polyethylene outer surface with a nylon lining. In order to determine the stiffness of this cross-section, it is necessary to know the second moment of area, I, for the section (2). This is not straightforward as there are two different materials involved. However, it can be shown (64) that it is possible to convert the two materials into an equivalent section as illustrated in Fig. 13(b). This equivalent section in the upper material will be of a slim 'T' profile with the web thickness given by

$$b_1^1 = \left(\frac{E_2}{E_1}\right) b_2$$

Alternatively it is possible to convert the section into one consisting totally of the lower material. This is shown in Fig. 13(c). In this case the flange width is given by

$$b_2^1 = \left(\frac{E_1}{E_2}\right) b_1$$

In each case, E refers to the modulus of elasticity and the suffices 1 and 2 refer to the two different materials. Even if the sandwich consists of 3 or more materials, the same methodology for converting the section to just one material may be used. The method to obtain the second moment of area, I, for the section obtained in Fig. 13(c) is given in any standard Strength of Materials book. The important point to note is that in bending, the equivalent section will behave in exactly the same manner as the sandwich section.

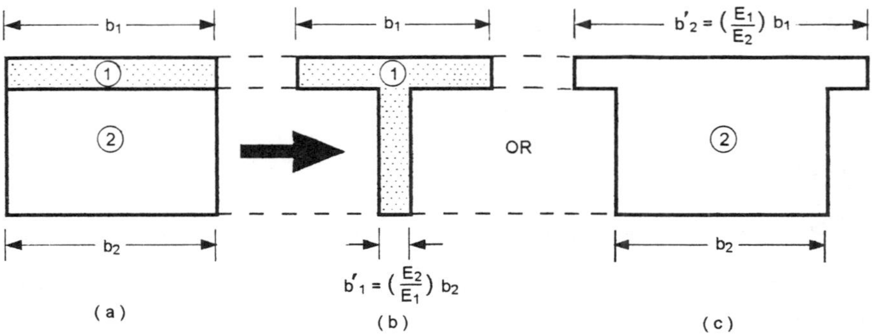

FIGURE 13 Equivalent sections

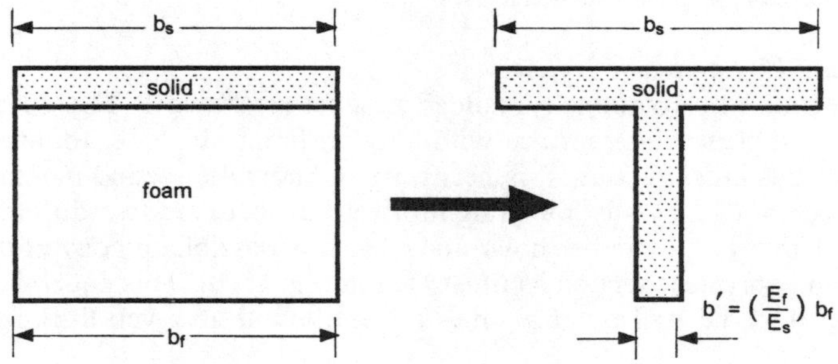

FIGURE 14 Solid/foam equivalent sections

Design of Foamed Sandwich Sections

As indicated above, it is now becoming common in the rotational moulding industry to have partially foamed sections, as shown in Fig. 14(a). In such cases, the design procedure to calculate flexural stiffness is to convert the composite structure to one consisting only of the solid material. This is shown in Fig. 14(b). The web width is once again given by

$$b^1 = (\frac{E_f}{E_s})\, b_f$$

where the subscripts refer to 'f-foam' and 's-solid'.

The only problem with eqn (3) is that although it is not difficult to get the modulus, E_s, for the solid material, it is not so easy to get a value for E_f. This is because there are a very large number of foam densities which are possible and it would not be feasible to expect the material supplier to provide data for all densities. However, it has been shown (65) that the modulus ratio is given by

$$(\frac{E_f}{E_s}) = (\frac{\rho_f}{\rho_s})^2$$

where ρ is the density of the material.

The use of this design approach is illustrated in the following Example.

Example - A rotationally moulded product has a foam sandwich structure as illustrated in Fig. 15(a). The 3 mm skin layers are solid polyethylene with a density of 930 kg/m^3 and the foamed core is polyethylene with a density of 100 kg/m^3. If a

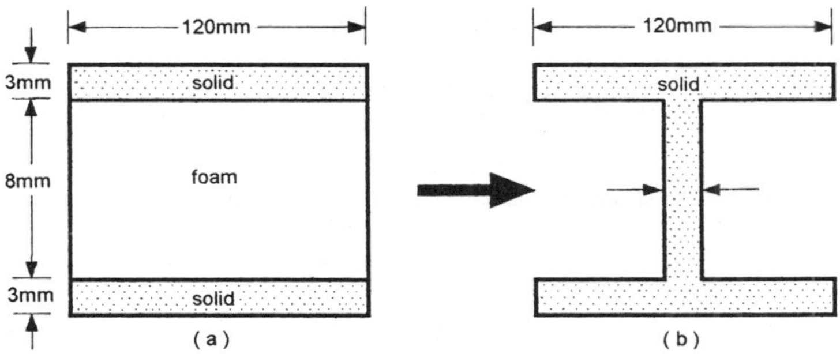

FIGURE 15 Foam sandwich section

panel of the moulding is 120 mm wide and is supported over a length of 500 mm, calculate

(i) the depth of a solid polyethylene beam which would have the same flexural rigidity
(ii) the weight saving by using the foam sandwich beam
(iii) the central deflection of the beam when it is subjected to a load of 20 N at mid-span.

The modulus of the solid polyethylene is 450 MN/m^2.

Solution

(i) The equivalent solid polyethylene section is shown in Fig.13(b). The width of the web is given by

$$b' = \frac{E_f}{E_s}(b) = \left(\frac{\rho_f}{\rho_s}\right)^2 b$$

$$b' = \left(\frac{100}{930}\right)^2 \text{x}120 = 1.39 \text{ mm}$$

The second moment of area, I, for the solid section will be given by (3),

$$I = \frac{12 \text{ x } 14^3}{12} - \frac{12 \text{ x } 8^3}{12} + \frac{1.39 \text{ x } 8^3}{12}$$

$$I = 22380 \text{ mm}^4$$

For a solid rectangular section to behave in the same way in flexure, it would need to have the same I value. Hence for the solid section

$$I = 22380 = (1/12)\ (\text{width x depth}^3)$$

$$\text{depth}^3 = \frac{12 \text{ x } 22380}{120}$$

(ii) The weight of the solid section (per m length) is given by

Weight (solid) = 120 x 13.1 x 10^{-6} x 930 = 1.46 kg

Similarly, for the original foamed section,

Weight (sandwich) = 2(120 x 3 x 10^{-6} x 930) + (120 x 8 x 10^{-6} x 100)

= 0.76 kg

Hence the foam section is 48% lighter than the equivalent solid section and yet they will both do exactly the same job.

(iii) When the beam is simply supported over a length of 500 mm, its central deflection, δ, is given by the following expression (64).

$$\delta = \frac{WL^3}{48EI}$$

$$\delta = \frac{20 \text{ x } (500)^3}{48 \text{ x } 450 \text{ x } 22380}$$

$$= 5.2 \text{ mm}$$

It should be noted that in this illustrative example, the modulus of the plastic was taken as a constant (450 MN/m^2). However, in real situations the modulus of a plastic decreases with time due to creep. Hence if one is attempting to predict the long-term behaviour of a moulding, it is very important to use the appropriate long-term modulus.

Selecting Plastics to Avoid Excessive Deformations

In a more practical design situation it is likely that a choice of material would have to be made. In order to compare the suitability of a range of plastics a series of calculations similar to those illustrated above could be carried out using time-dependent moduli from Table 1.

For example, it is evident that a thinner beam made from acetal could satisfy the design requirement due to the higher stiffness of acetal. But acetal is more expensive than polypropylene so would it be a more economical design? To resolve this

TABLE 2 MATERIAL DESIRABILITY FACTORS FOR ENGINEERING COMPONENTS

Component	Stress Limited	Deformation Limited
Rectangular beams of fixed width in bending	$\frac{\sigma_L^{1/2}}{\rho C}$	$\frac{E^{1/3}}{\rho C}$
Tensile rods or short columns in compression	$\frac{\sigma_L}{\rho C}$	$\frac{E}{\rho C}$
Long rods in compression, limited by buckling	—	$\frac{E^{1/2}}{\rho C}$
Thin wall cylinders under internal pressure	$\frac{\sigma_L}{\rho C}$	—
Flat plates under pressure	$\frac{\sigma_L^{1/2}}{\rho C}$	$\frac{E^{1/3}}{\rho C}$
Thin wall shafts in torsion	$\frac{\tau_L}{\rho C}$	$\frac{G}{\rho C}$
Solid shafts in torsion	$\frac{\tau_L^{2/3}}{\rho C}$	$\frac{G^{1/2}}{\rho C}$

where σ_L, τ_L — limiting direct or shear stress
E, G — Young's and shear modulus
ρ — density
C — cost per unit weight

Material	Desirability Factor
LDPE	5.7
HDPE	5.5
Polypropylene	7.8
Acetal Copolymer	2.1
PMMA	3.8
Rigid PVC	9.63
Nylon 66	2.5
Glass reinforced Nylon 66	3.3

type of dilemma it is now common practice to use Desirability Factors to compare materials on a cost per unit property basis. Table 2 lists a selection of these factors for some common loading situations (61,62).

For the example involving the polypropylene beam and using the 1 year modulus values from Table 1 and cost and density values for the individual plastics, it is possible to compare the desirability factors of a range of plastics for this application.

On this basis, rigid PVC would seem to be best. However, this is only the first step in the design process. It is now necessary to consider other factors such as moulding/fabrication method, possible environmental attack, special features such as transparency, electrical properties and so on. Note that the same procedure could be used to differentiate between different grades of polyethylene. The detailed design of rotomoulded products is considered in more depth in Chapters 6 and 7.

Selecting Plastics Based on their Permeability

To an ever-increasing extent plastics are being used as containers and as such their barrier properties are very important. In most cases the plastic container is replacing metal or glass containers each of which have almost infinite barrier properties. Plastics on the other hand are susceptible to permeation by gases and liquids although the extent varies widely. It may be seen from Table 3 that EVOH, PVdC, PAN, PA and PET have good barrier properties whereas the olefins, polystyrene and polycarbonate are not so good, particularly against oxygen and CO_2. Polypropylene and high density polyethylene are, however, good barriers to water vapour.

TABLE 3 PERMEABILITY COEFFICIENT K FOR DIFFERENT POLYMERS

Polymer	Oxygen $\frac{cm^3.100\mu m}{m^2.24hr.bar}$	CO_2 $\frac{cm^3.100\mu m}{m^2.24\,hr.bar}$	water vapor $\frac{g.cm}{m^2.24\,hr.atm}$
EVOH	0.4	1.2	0.8
PVDC	0.4	1.2	0.8
PAN	4.3	12.2	24.0
PA	22.5	47.3	23.6
PET	39.4	78.8	15.8
OPET	19.7	47.3	7.9
PVC	39.4	118.2	11.8
OPVC	27.6	59.1	7.2
HDPE	433.4	1,182.0	2.0
PP	906.2	1,773.0	2.0
PS	1,379.0	3,152.0	27-44
PC	1,891.2	2,955.0	47.3

The reason why plastics are so popular as containers is that they are light, unbreakable and flexible. However, in most cases the designer when turning to plastics for a high performance container may be seeking a range of qualities such as impact resistance, low cost, mouldability, resilience, strength, optical clarity, printability as well as good barrier properties. Therefore it is unlikely that any single plastic will have all the necessary attributes, and the modern trend is to produce multi-layer materials. This is like an alloy where one of the good barrier materials is included to reduce permeation and other materials are used to reduce cost, import flexibility, etc. A typical 3-layer combination is PC/PET/PC and a typical 5-layer combination is PP/adhesive/EVOH/adhesive/PP.

The permeation constant k is given by

$$k = \frac{Qd}{Atp}$$

where	Q	=	volume of medium
	d	=	wall thickness
	t	=	exposure time
	A	=	area
	p	=	pressure difference.

For a multi-layer combination the permeability coefficient may be calculated from

$$\frac{1}{k} = \frac{1}{d} \sum \frac{d_i}{k_i}$$

REFERENCES

[1]. Mooney, P.J., 'An Analysis of the North American Rotational Molding Business', PCRS Report, New Canaan, CT(USA), 1995.

[2]. Barrett, J., 'Rotomoulding moves into mainstream', Eureka, Nov. 1988, p. 56-59.

[3]. Schwarts, S.S. and Goodman, S.H., 'Plastic Materials and Processes', Chapter [14] Van Nostrand Reinhold, New York, 1982.

[4]. Taylor, P., 'Rotomoulding', British Plastics and Rubber, Feb. 1986, p. 22-27.

[5]. Ramazzotti, D., 'Rotational Moulding', Ch. 4 in Plastic Product Design Handbook, ed. by E. Miller, Marcel Dekker, New York, 1983.

[6]. Anon, 'Rotational Moulding' Plastics and Rubber Weekly, Jan. 27, 1987, p. 10-18.

[7]. Saffert, R., 'PVC Slush Moulding for Car Dashboards', Symposium No. 5, 3rd Annual Meeting of Polymer Processing Society, Stuttgart, April 1987.

[8]. Crawford, R.J. "Rotational Moulding of Plastics", Progress in Rubber and Plastics Technology, Vol. 6, No. 1 (1990) pp 1-29.

[9]. Kliene, R., 'Polyethylene materials for rotational moulding', paper presented at BPF Rotomoulder Seminar, Telford, Sept. 1989.

[10]. Simonsson, E., 'Polyethylene materials for rotational moulding', Neste Seminar on Rotational Moulding, Birmingham, June 1987.

[11]. Tomo, D., 'Rotational Moulding of Polyethylene Powders', in Basic Principles of Rotational Moulding, edited by P.F. Bruins, Gordon and Breach, New York, 1971.

[12]. Dodge, P.T., 'Materials for Rotational Moulding', ARM Seminar on An Introduction to Rotational Moulding, Chicago, 1985.

[13] Carrow, G.E., 'Cross linkable Rotational Moulding High Density Polyethylene', SPE Antec, May 1972, p. 762-765.

[14]. Rees, R.L., "Cross linkable Polyethylene for Rotational Moulding", SPE Antec, May 1982, p 621-623.

[15]. Grabensetter, T., 'Plastisols in Rotational Moulding', ARM Rotational Moulding Seminar, New Jersey, Jan. 1981.

[16]. Braeung, J., 'Moulding PVC Plastisols', ARM Workshop, Toronto, Canada, Oct. (1990).
[17]. Scott, J.A., 'A Study of the Effect of Process Variables on the Properties of Rotationally Moulded Plastic Articles', Ph.D. Thesis, Queen's University, 1986.
[18]. Edwards, W.A. and Horvath, L.J., 'PVC - The Beginning', ARM Rotomoulding Seminar, Washington DC, October 1986.
[19]. Anon, 'Halar Fluoropolymers for Rotomoulding', Allied Corporation, New Jersey, 1981.
[20]. Miller, W.A., 'Rotomoulding and Rotolining with E-CTFE'. SPE Antec, March 1969, p.72-77.
[21]. Anon, "Nylon 6 for Rotational Moulding', Allied Corporation, New Jersey, 1986.
[22]. Foster, W.E., 'Now you can Rotomould Polycarbonate', Plastics Technology, Nov. 1970, p.47- 49.
[23]. Titus, J., 'Rotational Moulding of Plastic Powders', AMC Engineering Design Handbook No. 706-312, April 1975.
[24]. McDonagh, J.M., 'Rotational Casting of Acetal Copolymer', Soc. Plast. Eng., March 1969, p.35-42.
[25]. Anon, 'Rotomoulding Casebook', Plastics Design Forum, May 1981, p.6.
[26]. Anon, 'Rotomoulding has new values for foams and thermosets', Modern Plastics, Nov. 1979, p.6.
[27]. Throne, J.L., 'Reactive Rotational Moulding', Polym. Eng. Sci., Sept. 1980, p. 899-919.
[28]. Narkis, N., Puterman, M. and Boneh, H., 'Rotational Moulding of Thermosetting Three Phase Syntactic Foams', Polym. Eng., Sci., May 1982, p. 417-421.
[29]. Harkin-Jones, E., 'Rotomoulding of Nyrim', 1st Year Research Report, Queen's University, Oct. 1989.
[30]. Anon, 'Rotomoulding Library', Association of Rotational Molders, 435 N. Michigan Avenue, Chicago, USA 1989.
[31]. Ward, N.M., 'Rotational Moulding Machines of Today' Neste Rotomoulding Seminar, Birmingham, June 1987.
[32]. Taylor, A., 'Mould Design' BPF Seminar on Rotomoulding, Telford, Sept. 1989.
[33]. Nickerson, J.A., 'Rotational Moulding', Modern Plastics Encyclopedia, McGraw-Hill, New York 1969, p. 806-812.
[34]. Hickey, J., 'Moulding for Rotational Moulding', ARM Seminar on An Introduction to Injection Moulding, Chicago 1985.
[35]. Beall, G., 'Engineers Guide to Designing Rotationally Moulded Plastic Parts', ARM publications, Chicago 1985.
[36]. Arato, P., 'Designing Rotationally Moulded Parts', ARM Seminar on Rotational Moulding, Chicago 1985.
[37]. Bockmann, A.F., 'Rotational Moulding - Recent Developments in Machine Construction and Moulding Procedure', Kunststoffe, 61, 7 1971, p. 1-5.
[38]. Crawford, R.J., Scott, J.A. and Nugent, P., 'Manufacture of Large Plastic Products by Rotational Moulding', 5th IMC Conference, Belfast, Sept. 1988, p. 98-119.
[39]. .Brenner, W. and Vermes, R., 'Rotational Moulding Today: New Materials and Equipment', Materials Engineering 1968, p. 64-65.
[40]. Anon, 'Promising Prospect for Powder Processing', Modern Plastics, April 1963, p. 74-79.
[41]. Anon, 'Powder Moulding is starting to Move', Modern Plastics, July 1961, p. 80-83.
[42]. Anon, 'Rotational Moulding - a technique of great potential', British Plastics, July 1968, p. 57-60.
[43]. Anon, 'Rotational Casting', Plastics Today, Jan. 1963, p. 9-12.
[44]. Anon, 'Rotational Moulding - star in ascendant' Plastics and Rubber Weekly, Oct. 19 1985, p. 22.
[45]. Anon, 'Rotational Moulding gains ground', British Plastics, March 1965, p. 124-128.
[46]. Wittman, C.A., 'Rotational Moulding - bigger parts, faster production', Materials Engineering 70, 4 1969, p. 40-42.
[47]. Rao, M.A. and Throne, J.L., 'Principles of Rotational Moulding', Polym. Eng. Sci., 12, No. 4 1972, p. 237-264.
[48]. Rao, M.A. and Throne, J.L., 'Theory of Rotational Moulding', SPE Antec, May 1972, p. 752- 761.

[49]. Throne, J.L., 'Some Factors affecting Cooling Rates of Rotationally Moulded Parts', Polym. Eng. Sci., 12, No. 5 1972, p. 335-339.
[50]. Throne, J.L., 'Rotational Moulding Heat Transfer - An Update', Polym. Eng. Sci. 16, No. 4 1976, p. 257-264.
[51]. Ribe, F.S., Progelhof, R.C. and Throne, J.L., 'Rotational Moulding', SPE Antec, May 1986, p. 20-26.

[52]. Crawford, R.J. and Scott, J.A., 'An experimental study of heat transfer during rotational mouldiug', Plast. Rubb. Process Appl. 5 (3) 1985, 239-248.
[53]. Crawford, R.J. and Scott, J.A., 'A Fresh Look at the Rotational Moulding Process for Plastics', Proc. 2nd IMC Conference, Jordanstown 1985, p. 495-513.
[54]. Crawford, R.J. and Scott, J.A., 'The Formation and Removal of Gas Bubbles in Rotationally Moulded PE', Plast. Rubb. Process. Appl. 7 (2) 1987, p. 85-99.
[55]. Crawford, R.J. and Nugent, P.J., 'Computer Simulation of the Rotational Moulding Process for Plastics', Plast. Rubb. Process Appl., 11 (2) 1989 p. 107-124.
[56]. Crawford, R.J. and Nugent, P.J., 'A New Process Control System for Rotational Moulding', PRCPA (in Press) 1992.
[57]. Crawford, R.J. and Nugent, P.J., 'Impact Strength of Rotationally Moulded Products', PRCPA (in Press) 1992.
[58]. Iwakura, K., Ohta, Y., Chen, C.H. and White, J.L., 'Experimental Investigation of Rotational Moulding', Int. Polym. Process, IV 1989, p. 163-171.
[59]. Crawford, R. J., 'RotoSoft - Computer Simulation of Rotational Moulding', Queen's University, Belfast.
[60]. Rabinovitz, E., and Rigbi, Z., 'Rotational Moulding of Polyurethane', Plast. Rubb. Process. Appl., 594, 1985, p. 365-368.
[61]. Crawford, R.I., 'Plastics Engineering' 2nd edition, Pergamon, Oxford (1987).
[62]. Crawford, R.J., 'Plastics and Rubber: Engineering Design and Applications', MEP Ltd., London (1985).
[63]. Crawford, R.J., 'Engineering Plastics', Kempes Engineering Yearbook, Chapter C1 (1991).
[64]. Benham, P.P., Crawford, R. J. and Armstrong, C.G. 'Mechanics of Engineering Materials', Longman (1996).
[65]. Moore, D.R., Couzens, K.H. and Iremonger, M.J., 'Deformation Behaviour of Foamed Thermoplastics', J.Cellular Plastics, May 1974, p.135-139.
[66]. Graham, B. 'Environmental Stress Cracking of Rotationally Moulded PE', Rotation, vol. 3, issue 2, 1994, p. 16-21.

ABBREVIATIONS

PVDC	-	polyvinylidene chloride
PAN	-	polyacrylonitrile
PA	-	polyamide ("nylon")
PET	-	polyethylene terepthlatate
OPET	-	oriented PET
PVC	-	polyvinyl chloride
OPVC	-	oriented PVC
PE	-	polyethylene
LDPE	-	low density polyethylene
LLDPE	-	linear low density polyethylene
MDPE	-	medium density polyethylene
HDPE	-	high density polyethylene
Super linear PE	-	metallocene based polyethylene
PP	-	polypropylene
PS	-	polystyrene
PC	-	polycarbonate
ABS	-	acrylonitrile-butadiene-styrene
PMMA	-	polymenthyl methacrylate (acrylic)
HIPS	-	high impact polystyrene

CHAPTER 2
Rotational Moulding of Polyethylene

R.I. Kliene

1. HISTORY OF DEVELOPMENT

The First Polymers

In the early part of this century the process of linking simple molecules of matter together to make long chain molecules was a process already known to researchers of organic chemistry.

The first successful industrial thermoplastic to be produced in any quantity was cellulose nitrate, discovered by Hyatt in 1868. It became the first substitute for natural materials, such as ivory and horn, and was used to replace them in piano keys, billiard balls and handles for cutlery.

Processing methods already in use in the rubber industry were readily adapted to this new material. Extrusion, compression moulding, callendering, granulating and eventually injection moulding were all processes that arose from these early developments, whilst the principles of blow moulding had earlier roots and descended from the glass blowing industry.

It was perhaps the high flammability of cellulose nitrate that led to the development of cellulose acetate and other cellulose derivatives, but by 1925 the first true synthetic plastics materials had arrived - polyvinyl chloride and polystyrene.

Early Rotational Moulding

The injection moulding process was by now in fairly common use and highly plasticised forms of polyvinyl chloride (PVC) which produced a mobile liquid, could be cast and cured to a solid by the simple application of heat. The casting of hollow objects by rotating viscous liquids in hollow moulds was a technique of even greater antiquity than blow moulding. Early Egyptians and Romans had cast hollow clay objects using this principle. In England the production of hollow lead castings was used in several industries including the manufacture of toy lead soldiers, and this principle is still used to produce chocolate eggs and novelty confectionery. The development of the rotational moulding process stemmed from these early ideas.

The Discovery of Polyethylene

In 1933 two researchers, Dr. R.O.Gibson and Dr. E.W.Fawcett, in their laboratory at the Alkali Division of Imperial Chemical Industries in Winnington, Cheshire, discovered that the gas ethylene, when subjected to high pressures in the presence of a catalyst, produced a solid substance with high molecular weight.

Previous attempts to produce long chain molecules from ethylene had only resulted in soft waxes, greases and oils. This momentous discovery was patented by ICI but for a number of years the polymer was developed mainly because of its unique electrical properties. They called this new material 'polythene' a term known in Britain almost as well as 'paper', or 'plastic', although the correct generic description 'polyethylene' is now used throughout the world. In effect ICI could have registered the name 'Polythene' for their particular product, but they missed this opportunity and registered the tradename of 'Alkathene', to link the discovery to their Alkali Division.

The Development of Polyethylene

Polythene or polyethylene's first applications were for dielectrics, and the Second World War's development of radar would not have been so dramatically successful but for its use. It is said that the early warning radar system used by the British was instrumental in the decisive victory of the 'Battle of Britain'.

Wartime production requirements gave impetus to polyethylene's commercial development with production at ICI's plant in Northwich increasing from a few tonnes to over 1000 tonnes a year by the end of the war. By then work had begun into developing a wider range of grades to produce materials of differing molecular weight and density, within the limitations of the process. By now the production of polyethylene was expanding rapidly in the USA where both Du Pont and the Union Carbide and Carbon Corporation had taken out patent licences from ICI to increase wartime production needs (Ref.1).

Grades of lower molecular weight than those used in cable extrusion processes could be readily produced and were found to be ideally suited for injection moulding. By far the largest use of polyethylene, however, is in the manufacture of thin films for packaging which are manufactured by an extrusion process developed in the USA soon after the war.

By the middle 1950's a number of other major chemical companies had begun production of polyethylenes mainly from ICI licences, including Texas Eastman, Dow and Monsanto. Also around this time Ziegler in Germany and Phillips Petroleum in the USA had developed methods of producing polyethylenes of much higher density using a low pressure process quite unlike the very high pressure method patented by ICI.

In North America, Du Pont found that their Canadian offshoot Du Pont Canada Inc. would need its own production unit following the breakup of their association with ICI in Canadian Industries Limited. The Canadian market was, however, relatively small and a versatile system capable of producing a wide range of grades was needed. This resulted in the development of a unique new variation now known as 'linear polyethylene' because of its reduced branching in the low to medium den-sity range. Using a modification of the Ziegler process and the introduction of another monomer – butene – in relatively small proportions, they produced what is, in effect, a polyethylene/butene copolymer.

The introduction of linear polyethylenes, with vastly superior properties over the original high pressure materials, significantly expanded the potential for the rotational moulding process.

The First Polyethylene Rotational Mouldings

The first polyethylene rotational mouldings did not appear until the early 1950's and were produced from granules. Heat was applied to a mould rotated at high speed so that centrifugal forces could provide the pressure necessary to deform the granules once they had softened. The success of this method was limited, as only relatively low molecular weight polyethylenes would soften and flow enough; and these, it was found, could suffer from a curious phenomenon known as 'environmental stress cracking'. To reduce the casting forces required in working this material, easier fusion was needed particularly as the use of higher molecular weight grades was also necessary to improve the physical properties of the end product. A means of addressing these problems would be to reduce the granule size, and this led to the use of powdered materials which will be discussed later.

The powders that were eventually produced by grinding were more fluid in their behaviour in their unmelted state than granules. In addition no pressure or moulding forces were required to cause adequate fusion and it was therefore an obvious step to adapt PVC plastisol rotational moulding machines with low rotation speeds (5-20 rpm) to produce the first hollow polyethylene parts.

The static moulding of powders (Engel process) was also an adaptation of the 'slush moulding' of PVC plastisols, whilst dip coating methods, in common use with PVC, were also developed once suitable polyethylene powders became available.

2. POLYETHYLENE PROPERTIES

The polyethylene grades used in the rotational moulding process are chiefly classified by reference to their Density and Melt Index or Melt Flow Index (MFI).

2.1 Melt Flow Index

Definition

Melt flow index is a number that indicates the viscosity of a molten polymer at a particular temperature. In other words it is a measure of its melt viscosity at that temperature.

Measurement (ASTM D 1238)

The equipment on which MFI is measured is, in effect, an extrusion plastometer (Fig.1). The material to be tested is placed in a heated barrel (190°C) and a piston is

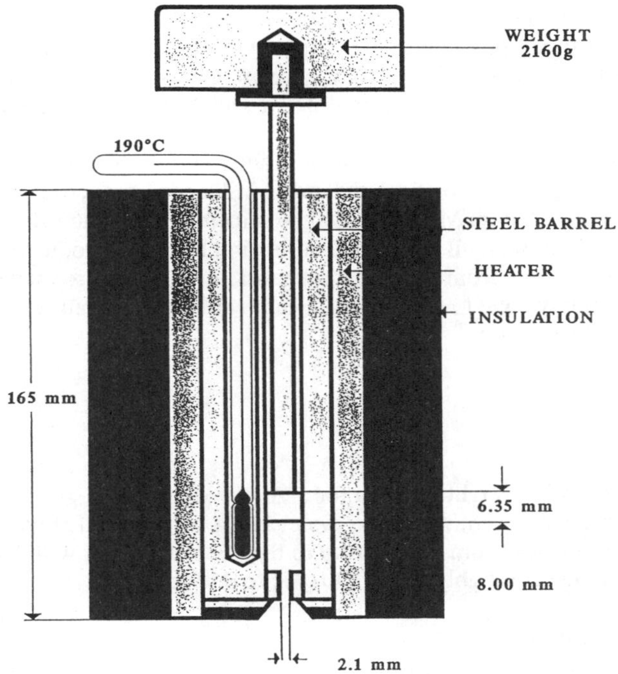

Illustration courtesy of Imperial Chemical Industries Ltd.

FIGURE 1 Apparatus for measuring melt flow index

mounted in the barrel so that there is a deadweight loading on the molten material which forces it through a die of specific length and diameter. The amount of material extruded through the die in ten minutes, measured in grammes, is defined as the Melt Flow Index or MFI.

There is a relationship between the MFI and the average molecular weight of polyethylene. The lower the MFI the higher is the average molecular weight and conversely the higher the MFI the lower is the average molecular weight.

The manufacturers of polyethylenes will provide technical data for the grades they market which will always include the Melt Flow Index.

Influence on Moulding

In broad terms the MFI is an indication of the viscosity of the resin when it has melted, and will help in determining the temperature and time cycle that will be necessary to reach optimum conditions. A high MFI, eg. 20, would demand a lower moulding temperature. If the temperature that has been set is too high or the process time too prolonged, the molten material may become overheated, resulting in excessive flow in the mould and the risk of thermal oxidation. The processor can reduce his moulding cycle time, thus improving output and reducing costs. A high MFI will also ease the flow of material into tight corners and produce parts with improved surface finish, both externally and internally, which may be of critical importance.

It could be argued therefore that a high MFI is desirable. Unfortunately, a resin with a high MFI will have poorer mechanical and stress influenced properties than those of resins of lower MFIs. A high MFI, it should be remembered, relates directly to a lower molecular weight, with all the risks that this implies. The product will be weaker in use. It might suffer from the phenomenon known as environmental stress cracking. It will certainly have lower impact strength at low temperatures and its chemical resistance will be poorer.

2.2 Density

Definition

The density of a substance is its weight or mass for a given volume. It is generally referred to as grammes per cubic centimetre. As water has a density of 1.0 g/cc it can be seen that all polyethylenes, from low density to high density types, will float in water, as their density range is roughly from 0.90 g/cc to 0.96 g/cc.

Measurement (ASTM D 1505)

There are several ways employed to measure the density of polyethylene but a common method used in many plastics laboratories is that which uses a density gradient column (Fig. 2). A glass cylinder is filled with a liquid made up of a series of alcohol/water mixtures in such a way that the most dense layer is at the bottom and each layer, of equal volume, decreases in density, with the lowest density layer at the top of the cylinder. Once the density gradient has been established it is calibrated from top to bottom by the use of a number of glass floats of known density. It is then possible to introduce pieces of the polyethylene and observe where they eventually settle and remain stationary in the column. The density can then be gauged as accurately as the number of calibrated divisions will allow.

The density of a polyethylene is governed by the closeness, or the packing together, of its long chain molecules. On the whole the molecules of polyethylene are largely similar and vary mainly in the number and size of branches that occur and the length of individual molecules. This similarity of form can allow close packing together of the molecules, whilst branching can locally prevent it. The closely packed areas or the orderly arranged zones result in formations known as ‘crystallites’, whilst the disordered areas are ‘amorphous’ (ie. the regions between the crystallites). Crystallites will form three dimensional shapes known commonly as ‘spherulites’ which grow in size and number during the cooling down of the melted material and its transition to a solid state. A polyethylene with closely packed crystallites and spherulites and less amorphous zones will have a higher

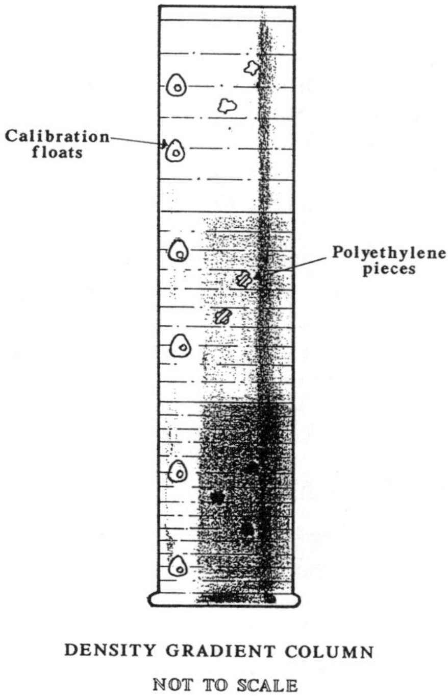

FIGURE 2 Apparatus for measuring density

density, and will be more crystalline. Materials with higher amorphous to crystalline ratios will have lower densities and will be less crystalline.

Influence on Mouldings

Most moulders will relate density to stiffness. The higher the density the greater will be the stiffness of the moulded part. It is also true that the higher the density the higher will be the thermal softening temperature, making the moulded product more suitable for high temperature applications. There is a negative factor. For a given MFI a higher density resin will have a lower impact strength than a lower density resin, and a lower 'environmental stress cracking' resistance. A higher density will also increase the risk of part failure due to notch sensitivity. This is a tendency to snap, accentuated when a part has been sharply scored and then bent. These properties can be improved by lowering the MFI.

2.3 The Density and Melt Flow Index Relationship

When selecting a polyethylene resin the choice of MFI is not one that should be made in isolation. The MFI should always be considered together with the density of the resin, and the use to which the product is being applied.

The choice left to the designer is usually a compromise. He should select a MFI low enough to provide good physical properties coupled with adequate ease of processing. To ignore this could be disastrous. A high MFI chosen for easy processing and good surface finish combined with a high density aimed at stiff, temperature resistant parts could result in very early failure in use. On the other hand if the moulding is not stressed in use and does not encounter conditions likely to be damaging then the advantages of this combination of properties are worth consideration.

Environmental Stress Cracking

Polyethylene mouldings can crack suddenly and cleanly when stressed either by external deformation or when moulded in such a way that there are significant internal stresses and when, at the same time, there are certain materials present that activate this process. A few examples are given in Table 1.

Table 1. Stress Cracking Agents

Stress Cracking Agents	Non Stress Cracking Agents
Acids Alcohols Detergents Hydrocarbons Organic esters Silicones Soaps	Inorganic Salts Mineral Oils Proteins Sugars Water

This phenomenon, known as 'Environmental Stress Cracking', is peculiar to polyethylenes in that these activators can be non-solvents and show no apparent absorption or chemical attack when failure occurs. However, some types of polyethylene are more susceptible to this type of failure than others; linear polyethylenes are in general far superior in this respect than the more highly branched high pressure types. The degree of stress that is present or applied also plays a part, and the fact that most rotational mouldings are largely stress-free stands the process in good stead compared to other polyethylene processing methods. The probability of cracking decreases with increased molecular weight (lower MFI) but it increases with increased density. Most polyethylene suppliers will provide data for their grades quoting 'hours to failure' when tested by ASTM D 1693, adapted from the original test procedure developed by the Bell Telephone Laboratory in the USA, who first discovered this phenomenon.

Materials with test results of 250 hours (to 50% sample failure) are good, whilst some grades tested to the maximum of 1000 hours without failure are excellent and virtually free of this problem altogether.

2.4 Other Physical Properties

Most manufacturers of polyethylenes used for rotational moulding produce a range of grades that will differ in density and melt index and can provide technical data covering their product range. As has already been discussed, many of the important characteristics, such as impact strength and environmental stress cracking, are density and melt index dependent. Some properties are not influenced to any large extent by these variations, for example the electrical or thermal characteristics, see Table 2.

Table 2. Polyethylene Electrical & Thermal Properties

Electrical Property	Value
Volume Resistivity	Approx. 10^{19} ohm cm
Surface Resistivity (75% RH)	4-5 x 10^{14} ohms
Permittivity (50 to 10^{10} c./sec)	2.28
Power Factor (tan δ)	Approx. 0.0001
Thermal Property	**Value**
Thermal Conductivity	0.0034 joule cm/cm sec °C
Specific Heat (solid)	0.55 cal/g
Specific Heat (molten 130 °C)	0.70 cal/g

Data courtesy of Imperial Chemical Industries Ltd.

Other properties are specifically related to density and not influenced greatly by the MFI; typically stiffness (or flexural modulus), surface hardness and the softening temperature, see Table 3.

3. POLYETHYLENE POWDERS

Process Powder Requirement

Like many other plastic materials, polyethylene is a poor conductor of heat. Yet it is necessary to convey sufficient heat to all the solid particles so that they can become molten and fuse together to form a solid mass. Clearly the rate at which the heat can penetrate to the centre of each particle will be dependent on particle size and volume; the smaller the particle the faster it will reach its fusion temperature.

Table 3. Polyethylene Physical Properties: Density Dependent

PROPERTY	UNIT	ASTM TEST	VALUE			
Density	g/cc	D1505	0.925	0.935	0.940	0.945
Flexural Modulus	MPa	D790	375	650	775	825
Tensile Strength at Yield	MPa @500 mm/min	D638	12	18	20	24
Surface Hardness	Shore D	D2240	52	57	59	61
Softening Point(VICAT)	°C	D1525	98	116	120	123
Approximate Linear Shrinkage	%	-	3	4	5	6

NOTE 1. - The data quoted are averages taken from various sources.

NOTE 2. - Other properties that are both density and MFI dependent cover too wide a range for inclusion in this publication. Data should be sought from technical data sheets issued by resin manufacturers.

result the pellets or granules of polyethylene are reduced in size by a system of grinding and sieving so that the average size of individual particles is about 300 microns with a maximum size of 400 to 500 microns for the bulk of commercial grades produced. The method of grinding, however, will create particles in a range of sizes from a fine dust up to 1mm or more in diameter. The size of the powdered particles, the distribution of particle size, the shape of the particles and their ability to flow freely when poured, will all have an effect on the efficiency of the process and the quality of the moulded part.

The Powder Grinding Process

The process used to reduce the polyethylene pellet or granule to a significantly smaller size has been called 'grinding' or 'milling' but neither of these descriptions truly describes what takes place in the grinding machine. Polyethylene is a relatively tough material. A normal milling or grinding process would simply result in shredding and tearing the particles in such a way that the resultant shape of each particle would be totally unsuited to subsequent processing requirements. The method that has been developed is perhaps more accurately described as disintegration by a series of high speed cuts.

The granules of polyethylene are fed into the centre of two plates, each with a series of radially arranged cutting edges as shown in Fig. 3. Usually one plate is held stationary while the other is rotated at high speed. The gap between the cutting edges of the two plates is narrower at their peripheries than at the centre. Any indi-

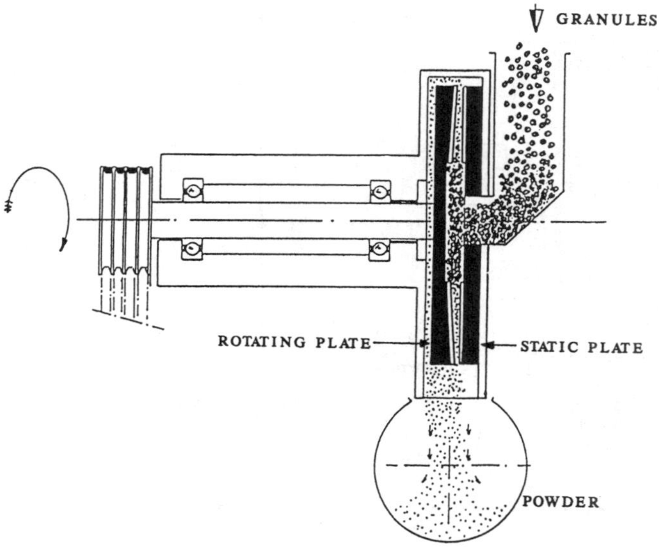

Illustration courtesy of Wedco Holland BV.

FIGURE 3 The grinding head

vidual granule is subjected to a cutting action which reduces its original size. The smaller sized particle is carried by centrifugal force outwards to a narrower section between the two cutting faces, where its size is further reduced. This sequence of high speed cutting and size reduction continues until all the particles, now much smaller, are ejected from the gap at the periphery of the circular cutting plates.

In this process frictional heat increases the temperature of the metal cutting faces, the individual polyethylene particles and the surrounding air. As the temperature rises so the particles of polyethylene will soften, making subsequent cuts cleaner and thereby reducing any tendency to tear and shred the granules. However, the temperature must be controlled so that it does not rise beyond the melting point of the polyethylene or indeed to a critical softening temperature before melting when the particles begin to adhere to each other and block the passage of new material entering the mill. There is a narrow 'window' in which to operate successfully, between the highest temperature the polyethylene grade can stand before sticking begins and a temperature below which shredding takes place which will produce particles of unacceptable shape and where some fibrillation has occurred.

Once the disintegrated particles have passed through the gap they fall into an air stream which conducts them to a sieve or series of sieves through which the finer particles will fall and be collected for use. Oversize particles held on the sieve are

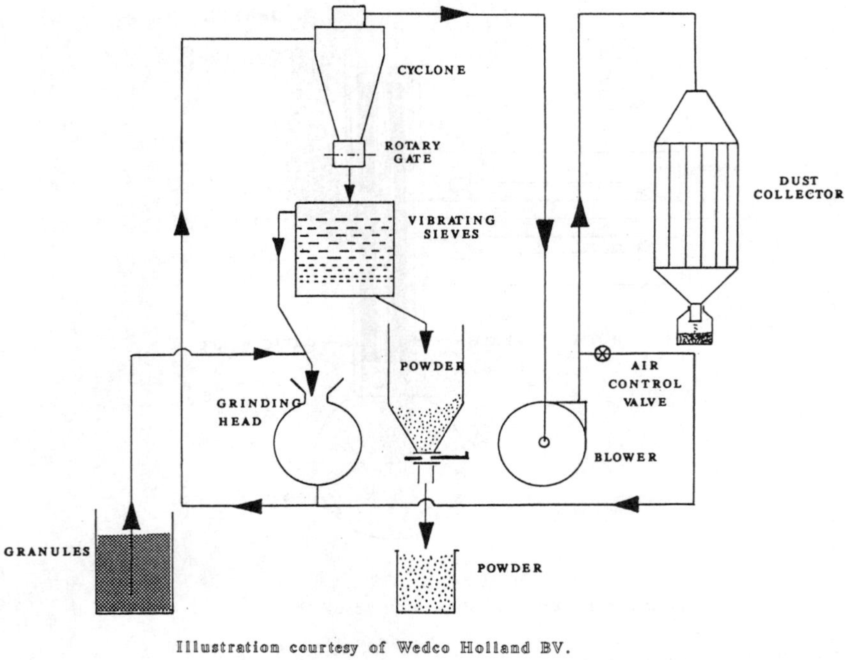

FIGURE 4 The grinding system

conveyed back to the mill for further disintegration (Fig. 4).

The particles of powder, viewed under a microscope, will be seen to be irregular in shape (Fig.5). The sharpness and condition of the cutting edges of the plates and the temperature of the particles at the time of cutting will have a significant bearing on the quality of the powder. The irregular shape of the particles is inevitable and can in fact be beneficial in the subsequent rotational casting process. Uniform spherical particles such as those produced in powder form from some reactor manufacturing processes can roll too easily before melting takes place and can result in uneven distribution particularly where complex mould shapes are required. A degree of particle irregularity reduces the chances of this happening. However, fibrillation or 'hairiness' of the powder caused by a tearing action can reduce its free flow properties and must be avoided.

Different types of polyethylenes and different grades within a type will each require different grinding parameters to obtain the optimum shape. As one would expect, tougher grades, ie those with higher impact strength, will be more difficult to cut cleanly in the grinding mill. Nevertheless a particular standard must be attained if the moulder is to achieve a satisfactory end result. Apart from particle shape there will also be an optimum particle size and particle size distribution

GOOD : Clean cut particles

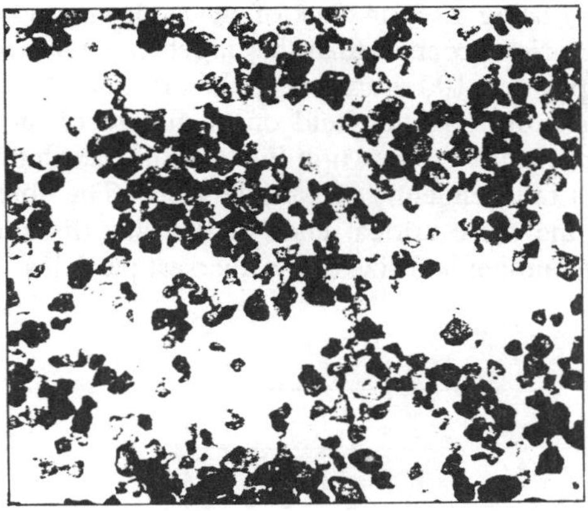

BAD : Shredded particles showing fibrillation

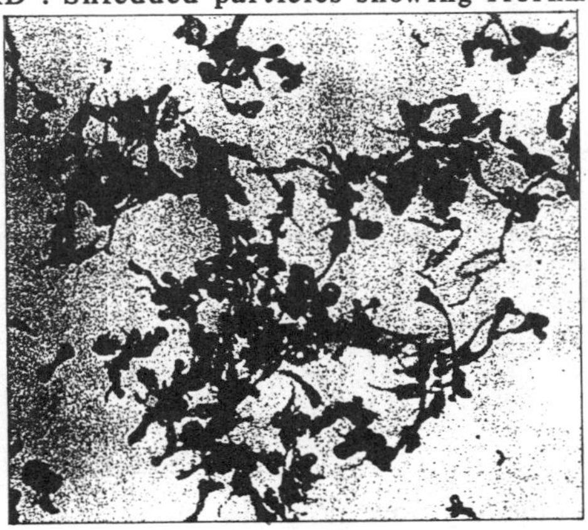

Magnification X5

Illustrations courtesy of Du Pont Canada Inc.

FIGURE 5 Ground polyethylene powder

passing through the sieves fitted in the grinder will range in size from a fine dust to a maximum size allowed by the coarseness of the sieve, usually about 500 microns.

Powder Quality

The quality of the powder produced should be monitored by using three simple tests.

1. Particle Size and Distribution Test (ASTM D 1921)
A particle size analysis is carried out using a set or stack of sieves of various mesh sizes ranging from 150 microns to a sieve coarser than the target maximum size - usually 500 microns. A sample of powder, 100 grams, is shaken through the stack of sieves for up to ten minutes and the quantity held on each size of sieve is weighed. The distribution of sizes is usually as shown in Fig. 6. This distribution is fairly typical and is achieved almost independently of particle shape. The object of the test is to observe and control the more critical ends of the size distribution curve, ie. the dust or 'fines', and the number and size of the coarsest particles.

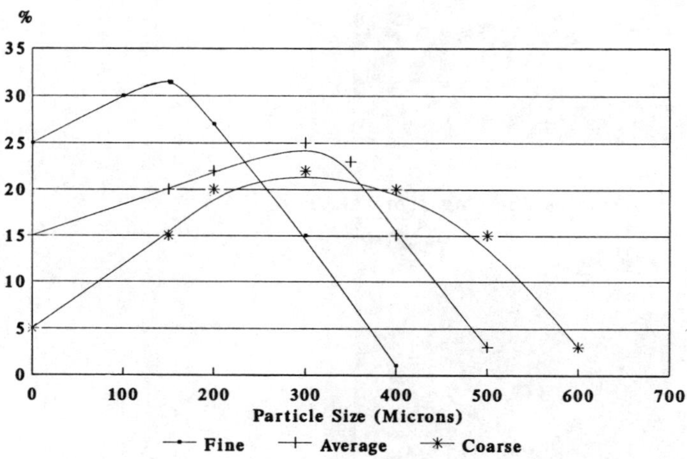

FIGURE 6 Particle size distribution

2. Dry Flow Test (ASTM D 1895)
The shape of the particles will affect the 'dry flow' properties of the powder. This is the rate at which the powder will flow through a funnel of specified shape and size (Fig. 7). A quality powder will flow through this funnel evenly and steadily much like sand pouring through an hour glass. Fibrillation or a tendency to 'hairiness' of the particles will slow this process and can, in extreme cases, stop the flow altogether. A 100 gram sample should flow through the funnel in about 30 seconds but more importantly the flow should be steady and uninterrupted.

3. Bulk Density Test (ASTM D 1895)
The bulk density of a powder is the weight of powder that is held in a given volume, without packing it in under pressure. The material falling through the Dry Flow funnel is caught in a receptacle of fixed size and weighed (Fig. 7). A fibrillated powder occupies more space than a cleanly cut powder (Fig.8) and indeed it can

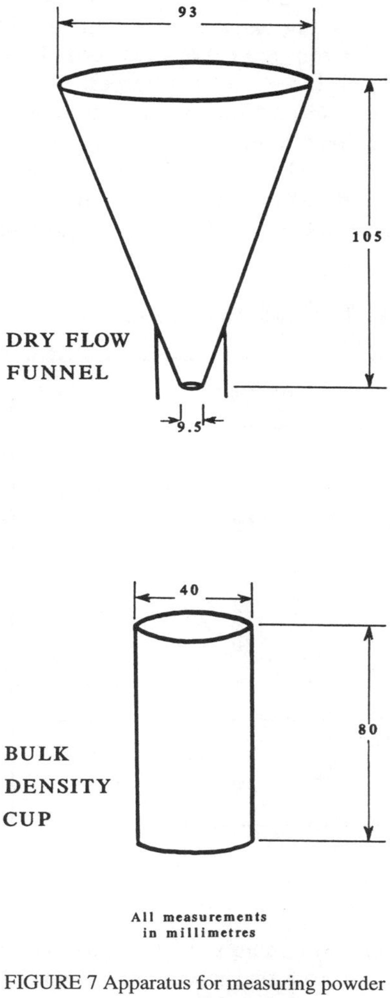

FIGURE 7 Apparatus for measuring powder flow and powder bulk density

FIGURE 8 Effect of powder grind quality on bulk density

be demonstrated that a low bulk density (low weight per volume) can be the result of poor grinding practice. However, it is not always as good a guide to quality as the 'Dry Flow' test, as abnormal particle size and size distribution will also have an effect. Nevertheless it is another method of quality assessment in common use.

Powder Characteristics and their Effect on Moulding

Powders can be ground to the needs of individual processors and depending on the melting rate, can range from fine to coarse. The desirability of varying the average particle size can only be assessed when the moulding parameters are known. Fine powders with a maximum particle size of 400 microns are not always needed. Most polyethylene grades in the normal range of melt index will mould perfectly well with a maximum particle size of 500 microns, and materials with very high MFI and low density will flow out smoothly with even higher particle sizes (ie. 600 microns or more). The finer powders tend to have slower dry flow properties but will produce good surface detail (Table 4). On the other hand, coarse powders will need more heating to achieve similar surfaces if not compensated by a higher melt flow index.

Table 4. Powder Analysis, Typical Properties

Microns	Fine	Average	Coarse
600	-	-	3
500	-	3	15
425	1	15	20
300	15	25	22
212	27	22	20
150	32	20	15
PAN	25	15	5
Bulk Density (g/cc)	0.30	0.35	0.40
Dry Flow (seconds)	35	30	25

Particle Size

It is reasonable to expect a fine powder to melt more rapidly than a coarse powder of the same grade of polyethylene, and it is also true that in any given sample of powder there will be powder particles of all sizes varying from dust to the largest size. The presence of dust or at least a reasonable proportion of fine particles can improve both the melting rate and the surface finish of the moulding.

In the rotational moulding process a proportion of the powder in the mould tumbles over itself as the mould is slowly rotated. This tumbling action, together with some inevitable vibration, results in a natural sieving of finer particles between the coarser particles so that a degree of layering takes place before the particles reach their melt temperature (Fig.9). Finer particles will filter downwards to the mould surface and in due course melt and adhere to it, while coarser particles tend to rise

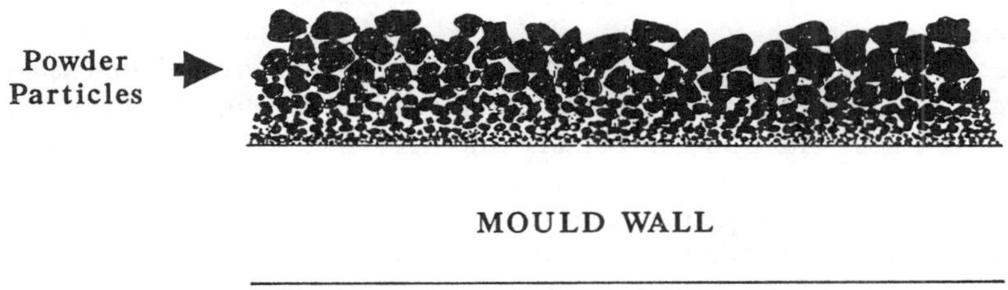

FIGURE 9 Filtration – powder particle size layering

and eventually melt at the inner surface of the moulding. As the powder is a mixture of solid particles and air, some trapping of air pockets will occur as the powder melts. The amount of air that is trapped between the particles, and the size and number of these air pockets, are affected by the powder particle size and distribution, and the quality of the powder grind. The air pockets that are trapped between fine powder particles are naturally smaller than those trapped between larger particles. Since the powder layer nearest to the mould surface is usually composed of finer particles or dust, the majority of bubbles in this layer will be smaller than those in the melt. These small surface bubbles will more readily disappear as the skin of the moulding is formed. A powder completely devoid of fine particles can result in voids at the mould/melt interface that are large enough to remain through the cycle of heating and cooling and produce mouldings which display a pitted surface. The ease with which bubbles will disappear from the surface will also be affected by the melt viscosity of the resin. Whilst a proportion of fine particles is desirable, too large a percentage of fine powder usually results in an increased proportion of neighbouring size material with the effect of moving the average particle size downwards. This would result in faster melting throughout, and, if not controlled, rapid thermal oxidation. The moulder needs to be assured that for a given grade of polyethylene in use the cycle times he has established will be reasonably constant and it is therefore important that the sizes of particles and their distribution are also constant, or at least within the limits that can reasonably be achieved by the grinding process.

Particle Shape

A poorly ground powder is usually one where some shredding or tearing has occurred resulting in a degree of fibrillation. This can clearly be seen at microscope magnifications as low as x10 or x20. A particle may have a 'fibrous tail' which can be two or three times as long as the particle itself. Large amounts of particles of this type will prevent the powder from flowing freely. The presence of 'tails' or 'hairs' results in a type of moulding fault known as 'bridging'. Instead of flowing into a narrow recess in the mould cavity the particles are held back, unable to enter possibly locked by intertwined hairs, where they melt and then adhere to other particles which cannot pass by. A 'bridge' is thus formed (Fig. 10) and remains as an imperfection on the inside surface of the moulded part and may even prevent complete fusion of the outer surface. Clearly, because of the increased dimension of the particle and its possible enlargement by attachment to other particles before melting, it can make the filling of corners with tight radii impossible. In addition the irregular shapes of several combined particles will in themselves entrap more air pockets and cause an increased level of voids and enlarged bubbles in the melt. It is not surprising that mouldings containing bubbles of increased dimensions and frequency in the moulded wall will have significantly lower impact strength, tensile strength and elongation. In fact all physical properties will be impaired.

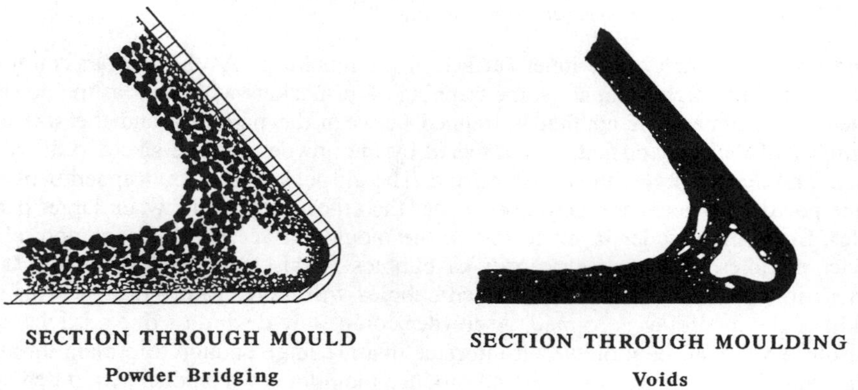

FIGURE 10 Bridging of poor quality powder

Even in less extreme cases where the particle shape is not ideal, but adequate, there will be a tendency for a larger than average number of bubbles to form. An increased heating time might be enough to improve the visual effect but this in turn can result in a greater risk of oxidation and resultant reduction of end properties.

Finally, another effect may make its appearance. Fibrous particles can collect into fluffy 'balls' rolling about on the top surface of the powder when tumbling and distribution within the mould is still in progress. When all the distributed powder has melted to form the wall of the moulding these 'balls' will eventually become anchored to the melted inner surface and form large irregular shaped lumps sometimes referred to as 'scrambled eggs' because of their appearance (see Fig.11). When this has occurred the moulding cannot be improved by extending the heating cycle in the hope that the unevenness will flow out. The internal surface may become glossy or shiny but the underlying irregularity will remain.

4. POLYETHYLENE ADDITIVES

Rotomouldable polyethylenes can contain a variety of additives which may be incorporated either during their manufacture or subsequently added to meet the needs of specific applications. As supplied, most rotational moulding grades will

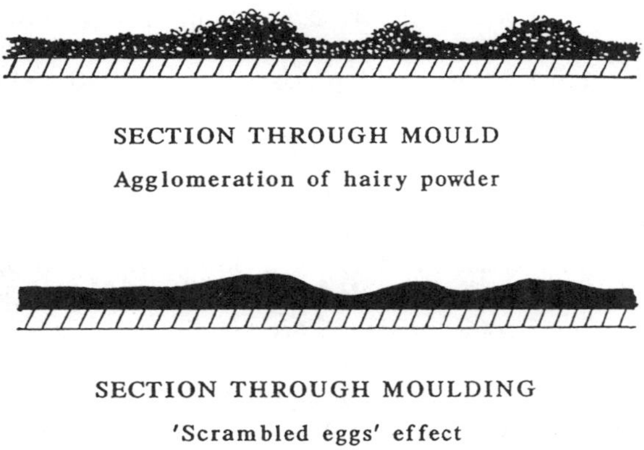

FIGURE 11 Effect of agglomeration of poor quality powder particles – 'scrambled eggs effect'

contain antioxidants to protect the material during long moulding cycles and in subsequent use where high service temperatures may be encountered. In addition, since many mouldings are used outdoors in direct exposure to UV light, UV stabilizers are usually present. However, the resins are otherwise provided in their natural state, which is in the form of granules with a cloudy, translucent white appearance.

Pigmentation

To produce coloured mouldings, pigments can be added. It should be remembered that pigments are very finely divided solid particles of organic or inorganic materials which remain discrete from their resin host and as such will have some effect on the physical properties of the moulded part. For this reason the level of addition, or pigment concentration, should be kept to a minimum. There are two methods of pigment addition, these being the addition by 'melt compounding' or by 'dry blending'.

Melt Compounding

Colouring of injection moulded or blow moulded parts is usually achieved by way of pigment melt compounding, the process taking place during extrusion of the molten polymer. Polyethylenes used in rotational moulding can be coloured using this method if appropriate extrusion equipment is used prior to grinding.

The chosen rotational moulding grade in granule form is fed to a heated extruder, and the pigment, usually in the form of a 'masterbatch' or colour concentrate, incorporating several pigments, is blended with it so that the product leaving the extruder is the required shade and intensity of colour.

It should be remembered that polyethylenes can suffer from environmental stress cracking and the presence of pigment may reduce resistance to this phenomenon. However, it is not always realised that the carrier for the pigment in the masterbatch or concentrate can also be a potential hazard even though it is present in small amounts. The common practice of using waxes or high MFI polyethylenes should be avoided and it is best to use a carrier with properties close to those of the chosen grade if not the grade itself.

Dry Blending

Dry blending involves the physical blending together of pigment powders and the natural polyethylene ground powder either by tumbling them together or using a high speed mixer. The latter produces a better result due to more efficient mixing. Either method is not as costly as melt compounding. Since the dispersion achieved using this method is not as good as that achieved when melt compounding, caution must always be exercised in controlling the amount of pigment added.

There are occasions when dispersing agents and pigment carriers are added to improve dispersion and pigment handling. This should only be done after assessing the risk of reduced physical properties in the moulded part.

Choice of Colouring Method

Melt compounding is more expensive but results in parts with greater colour intensity due to a high degree of dispersion of the pigment particles and even distribution throughout the moulded part. Because the dispersion achieved with modern compounding equipment is generally of a high order it is possible to reduce the amount of pigment needed to produce a particular result. There is, however, the added cost of compounding to consider. Where costs are critical there is a tendency to use the dry blending method.

Melt compounding polyethylene ensures that the pigment used has been encapsulated within the resin and that it will remain there in a uniformly dispersed condition. The tumbling of the powder within the mould during its rotation cannot disturb the distribution of the pigment nor can it migrate to the surface of the moulding. This can happen with dry blended powder, particularly if the pigment addition is too high. At above 0.25% concentration, migration of the pigment can occur, and a pigment-rich layer may form on the inner surface of the moulding. In some cases it is possible to wipe pure pigment off the inner surface of the moulding; and a good test is to rub the surface with a piece of cotton wool.

The presence of loose pigments is not desirable for several good reasons:

1. Some pigments may be considered suitable for use in contact with food or potable liquids but this assumes that the pigment is melt encapsulated. This is not the case with dry blends. Clearly food or drink containers coloured by this method may contaminate their contents.
2. The moulding of toys must raise questions about safety where children are concerned.
3. In time pigment will stain the mould surface and this could become a problem if mouldings of different colours are required from the same mould.
4. A separate area is needed for the blending process in order to confine the

pigment powders and prevent them from contaminating other work areas or finished parts. Operators working with dry pigments should wear masks and be equipped with suitable clothing to avoid health hazards.

Perhaps the greatest problem with the dry blending process is the effect on the physical properties of the moulding. Several studies have been carried out over the past few years and in general the results are much the same. At pigment concentrations above 0.25% there will be some deterioration of impact strength. If the blending of the powders is not efficient a similar deterioration of properties can occur at lower levels, probably due to the presence of pigment agglomerates which provide stress concentration points in the moulded part. In one of the studies (Ref.2) which compared both colouring methods, rotational mouldings were pigmented at different levels and then impacted by dropping the mouldings at -25°C and measuring the height at which failure occurred. The results are shown in Fig.12. From this it can be seen that there is a significant reduction of impact strength at all levels of dry colouring and that the impact resistance at 0.25% pigment concentration is the equivalent of a melt compounded material with four times the level of pigment. Addition of pigment at this high concentration by melt compounding is not required and rarely exceeds 0.5% where the impact strength has not been significantly affected.

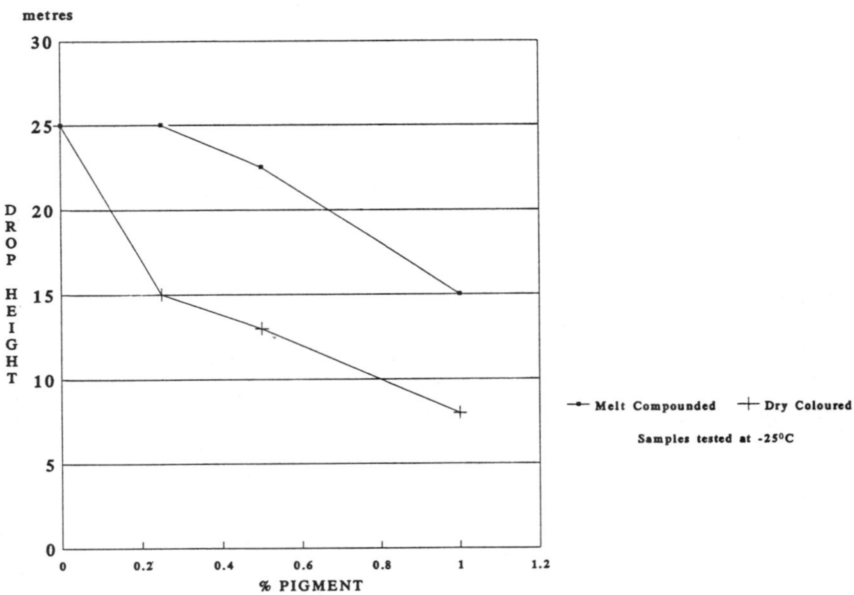

FIGURE 12 Effect of pigment on impact strength

Pigments

Pigments are broadly classified as organic or inorganic and are in the form of extremely fine powders. Typical examples of these types of pigments are shown in Table 5. The choice is largely governed by the required shade, density of colour, stability during processing and subsequent use and their cost.

Pigments with reasonable or good light and heat stability can be selected to achieve most shades of colours and once encapsulated in polyethylene can be suitable for use in contact with food or potable water.

Table 5. Pigment Types

ORGANIC Generally complex chemicals		INORGANIC Generally have a metal component	
Phthalocyanines	Green Blue	Cadmiums	Red Yellow Orange
Azo Pigments	Red Yellow Orange	Chromes	Yellow Orange
Quinacridones	Purple Violet	Titanium Dioxides	White
Carbon	Black	Iron Oxides	Brown Black
		Sodium Silicates	Ultra Marine Blue
		Cobalt	Blue
		Titanates	Ochre Yellow Brown

In general, organic pigments are more expensive than their inorganic counterparts but have greater colour strength. They are not usually as heat or light stable.

Compounds containing Carbon Black

As with pigments, carbon black can be added to polyethylene by the melt compounding technique. If a black colour is the simple requirement then the level at which carbon black is added need only be as high as 0.1%. However, the object of adding carbon black is generally for quite a different purpose. Carbon black is the best screen against the deleterious effect of UV light. Ultra violet rays from sunlight can catalyse an oxidation process, resulting in the rapid embrittlement of mouldings used outdoors.

Carbon black is a finely divided powder with an average particle size in the region of 20-60 millimicrons. When protection from UV light is required the degree of dispersion of this very fine powder is critical. The simple dry blending of the carbon black with unstabilized polyethylene powder is inadequate. Deterioration of the moulded part will only be marginally delayed. An efficient melt compounding process is necessary. This can be done by employing masterbatches or concentrates containing high levels of carbon black and where a high degree of dispersion has already been achieved in the base resin. This is then let down by a secondary melt compounding process which reduces the concentration from around 20-50% carbon black to 1-2%. Provided the secondary stage is an efficient system which distributes the concentrate uniformly and evenly, concentrations as low as 0.5% will provide useful protection for as much as 10 years outdoors. At 1.5% concentration the expected service life of a moulded part can exceed 20 years (see Fig. 13). The required degree of dispersion to achieve these results cannot be judged by the external appearance of mouldings. All mouldings, regardless of dispersion quality, will appear evenly black to the naked eye even in thin sections. A true assessment of carbon black dispersion and concentrate distribution can only be made by microscopic examination where streaking, smearing and carbon agglomerates can be seen. Specimens can be prepared by melting down pin head size pieces of the compound or by taking microtomed slices of the moulded part. The thin film made by these methods can then be viewed by transmitted light at magnifications of around 100.

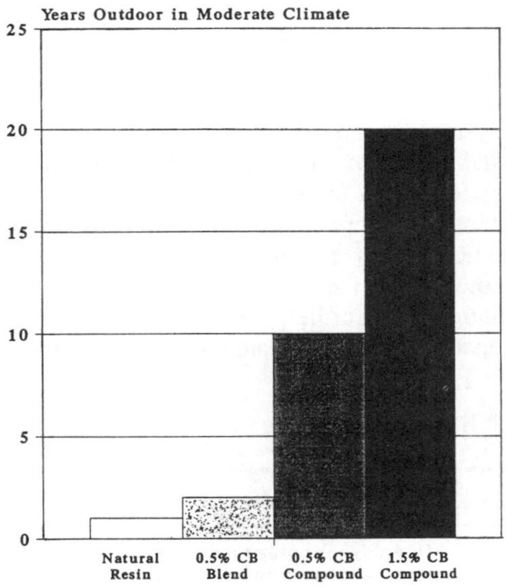

FIGURE 13 Effect of carbon black (CB) on weathering of polyethylene

It must be remembered, however, that a black surface will absorb more heat than white or natural surfaces, and in strong sunlight temperatures in the region of the melting point can be reached. This could drastically affect the usefulness of a polyethylene moulded part. In addition the black colour itself may have aesthetic disadvantages. The clear alternative is to use an effective UV stabilizer system where there can be a wide choice of colour options.

UV Stabilizers

Over the past two decades significant strides have been made in the development of UV stabilizers which inhibit the degradation of the polymer (Ref.3). There are three major classes of UV stabilizers now in use, see Table 6.

The first UV stabilizers used in polyethylenes for rotational moulding were UV absorbers (UVA). These chemical compounds absorb UV radiation and transform the UV energy into heat which is harmlessly dissipated through and from the polymer.

Another method of stabilization is to subdue or deactivate the excited state of the functional groups in the chain molecule, and again dissipate the accumulating energy as heat before photocatalytic oxidation can occur. These additives are known as quenchers. A disadvantage here is that the compounds are commonly nickel salts which produce an unacceptable green discolouration in thick sections. Consequently they are mainly used in thin film applications.

The third and most effective method is to incorporate a hindered amine light stabilizer (HALS). Although the precise mechanism of the process is not fully understood it is known that photoxidation is prevented by the scavenging of free radicals as soon as they are formed. To reduce migration and extraction of the stabilizer, polymeric forms of these hindered amine stabilizers were developed and have proved highly stable and extremely effective.

The effectiveness of the stabilizer systems is dependent on the concentration present but practical limits have been established for each type, dictated by its effectiveness, compatibility and acceptance for use in food contact applications. Whilst monomeric forms are extremely efficient stabilizers, many rotational moulded parts require long term protection in extreme outdoor conditions where the polymeric HALS are superior. In unpigmented mouldings the combination of HALS with a UVA boosts the resistance to UV radiation even further. Where inorganic pigments are added to provide colour this combination is not an advantage, but in the majority

Table 6. UV Stabilizer Types

TYPE	METHOD	COMMON COMMERCIAL COMPOUNDS
Absorbers	Reduce light penetration by absorbing the UV	Hydroxybenzophenones Hydroxyphenylbenzotriazoles
Quenchers	Deactivate and dissipate accumulated energy as harmless heat	Nickel-phenolates (Nickel salts)
Scavengers	Scavenge free radicals as soon as formed	Hindered amine light stabilizers (Polymeric HALS)

of cases the resins are supplied in the natural state with the combined package already present.

Some pigments can take over the role of the UVA and reduce the penetration of harmful UV radiation (Fig. 14). On the other hand certain pigments can act as sensitizers and reduce the effectiveness of the UV stabilizer.

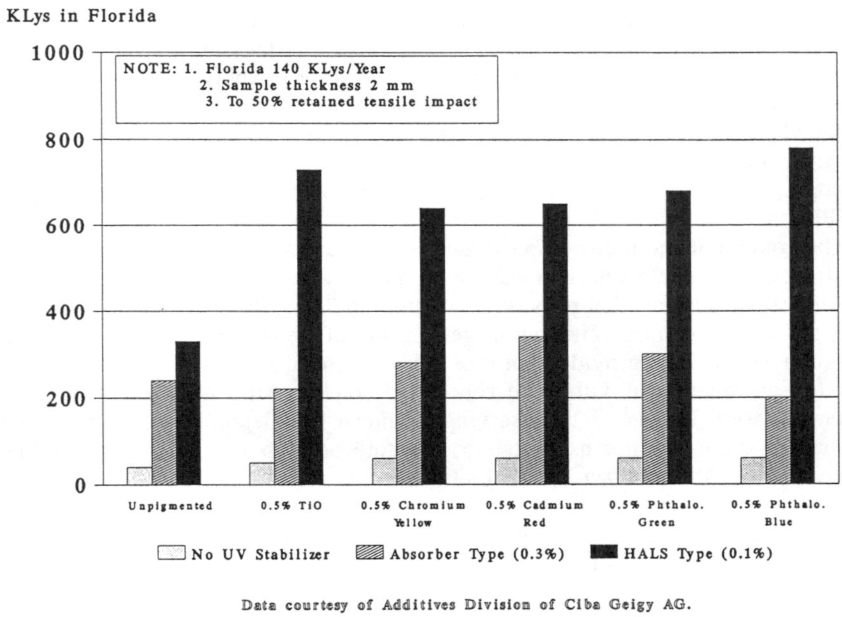

FIGURE 14 Influence of UV stabilizers and pigments on light stability of polyethylene

HALS compounds will scavenge oxidation promoting radicals formed either by UV radiation or thermal degradation and as a result contribute to long term heat stability. The long heating cycles of rotationally moulded parts could also have an influence on the long term UV protection. An efficient antioxidant system must be present to protect the polymer from immediate thermal degradation during processing; however, excessive heating could reduce the effectiveness of the UV light stabilizer. Lower density, high MFI grades of polyethylene which require less temperature or time in the processing oven could have significantly better outdoor service performance than higher density low MFI grades requiring more process heat.

It can be seen therefore that the outdoor service life of a rotational moulded part is dependent on a combination of both the antioxidant and UV stabilizer systems and that it can be affected by the presence of pigments as well as the thermal treatment it has received, whether caused by excessive heating or dictated by the physical characteristics of the polymer in use.

5. POLYETHYLENE PROCESSING

The Use of Release Agents

During the rotational moulding process heating of the polyethylene powder and subsequent melt is most extreme at the mould/polymer interface. At this interface a degree of surface oxidation can occur, resulting in a chemical bond between the polyethylene and the metal mould surface. In addition to this form of adhesion, polymers in their molten state can flow into mould surface imperfections, causing localized mechanical keying. It is necessary therefore to apply a heat stable barrier between the metal and the polymer to prevent adhesion and assist release of the moulded part. These barrier materials are known as release agents.

Many different chemical types of release agent exist including soaps, stearates, silicones, siloxanes, silazanes and fluorocarbons e.g polytetrafluoroethylene (PTFE).

The rotational moulding process places severe demands on the release agent coating. Considerable research has been undertaken in developing high performance release systems that provide rotational moulders with durable high temperature resistant coatings. The latest generation of semi-permanent products are favoured since they provide multiple releases from a single coating, are non-transferring (a problem with the earlier stress cracking silicones) and do not build up on the mould surface, thus preserving mould surface detail. In addition the semi-permanent types are not as prone to releasing at high part temperatures ('pre-release'), a further characteristic of silicones which can lead to part distortion on cooling.

Where 'pre-release' is not considered a problem and the handling and control of semi-permanent products is, PTFE and equivalent materials have been used. These will provide a greater release capability but are expensive to apply and refurbish when they eventually break down.

The semi-permanent types of release agent are most commonly used and are in the form of a solvent carried polymeric resin which coats the metal surface with a continuous film. Various systems are available, and types can be chosen to provide higher or lower temperature release with more or less lubrication or 'slip' to ease the moulding from the mould. They can be sprayed onto the mould surface or be wiped on with clean cotton cloths. In all cases the mould must be clean and free of contaminants such as pigments, oxidised polymer or rust to achieve the best results.

The Effect of Heating

With the best quality of grind that can be made there will always be air trapped between the powder particles during melting. As heating continues these bubbles eventually disappear. It is known that the bubbles that have been formed do not travel through the melt to the inner surface. The nature of their disappearance has been explored by several researchers. The original explanation offered by work carried out by P.Y.Kelly (Ref.2) provides a perfectly satisfactory conclusion in this writer's opinion. Kelly's work showed that the bubbles once formed did not move any distance in the melt. The viscosity of the molten polyethylene is such that the bubbles do not have sufficient buoyancy to rise to the surface. They shrink in size and eventually disappear. It appears that the air dissolves into the polymer. At a melt temperature of around 150°C a collection of bubbles of different sizes will have formed but by the time the melt temperature has reached just over 200°C even the largest bubbles disappear. From a moulder's viewpoint the most important conclusion is that, given a powder of reasonable quality, he can expect to remove all or most of the bubbles that have formed by allowing sufficient time for the critical temperature to have been reached. Kelly's work was limited to one particular type of polyethylene of medium density and melt flow index. There is no reason to believe that his findings do not also apply to other resins, with only the critical time/temperature relationship being different.

Because the heating cycle for a rotational moulding can be lengthy, anywhere up to half an hour or more for a wall thickness of 10 mm, it is clear that the resin needs to be protected from thermal oxidation and resultant deterioration of material properties. As previously stated, antioxidants are normally added to rotational moulding grades which under normal processing conditions have proven effective. Excessive heating, whether deemed necessary or through inadequate process control, may totally consume the antioxidants present and expose the moulded part to the risk of subsequent failure in use due to rapid embrittlement. There are usually signs that excessive heating has occurred. The interior of the moulding, where the greatest oxidation occurs, can develop an acrid smell and a 'sticky' glossy surface. Obviously this should be avoided. If a lengthy cycle is required to produce heavier or thicker walled mouldings the temperature should be reduced. Shortening the cycle and increasing the temperature may result in a part where the cycle time is too short to allow the disappearance of bubbles, even though the surfaces of the moulding are overheated and on the point of oxidation. Because of the poor thermal conductivity of polyethylene it is possible to overheat a thick walled moulding and severely oxidize the surface skin whilst the inner material below the surface has not reached a temperature high enough for long enough for the bubbles to be removed. Clearly in all cases the melt flow index, and to some extent the density, will have a part to play. Lower MFI grades will require longer heating and higher temperature with a greater risk of degradation.

When polyethylene is heated to a temperature high enough to cause melting there is a transition from the crystalline solid state to the amorphous molten state where the crystal structure has totally collapsed, ie. a molecularly disordered state. This is what is happening when the first layers of powder begin to fuse to the mould surface and take on a transparent appearance. The heat conduction from layer to layer will in time result in a uniform molten skin. Once this has happened and a sufficient time then allowed for complete fusion, marked by the virtual disappearance of bubbles and a smooth inner surface, cooling can commence.

The Effect of Cooling

As the melt temperature drops from its peak at over 200°C the disordered molecules will strive to reform in a more orderly manner, resulting in the 'rebirth'

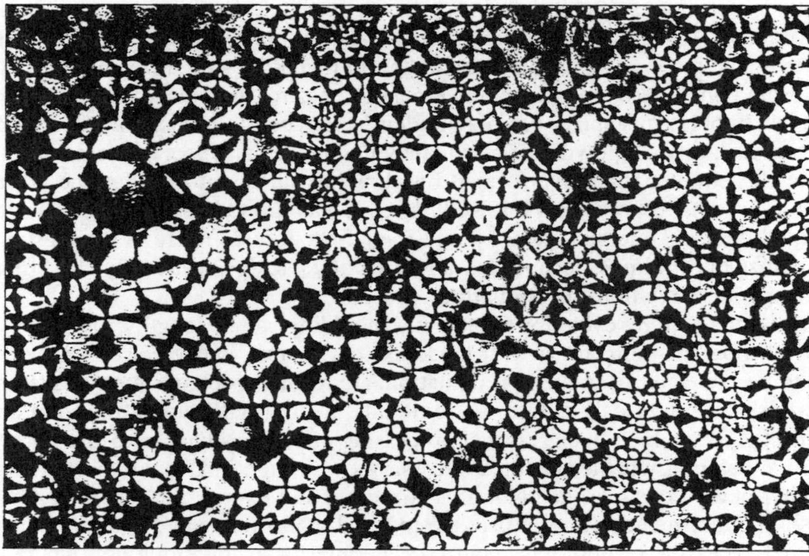

Illustration Courtesy of Du Pont Canada Inc.

FIGURE 15 Spherulite formation viewed with polarized light

of crystallites. These crystallites will continue to form and grow in size as cooling continues. The process takes time, and a fully reordered crystalline arrangement could take several hours to stabilize. Viewed under powerful magnification these changes can be seen occurring, and the neatness and the orderliness of the resultant spherulites can be seen in the solid material when viewed with polarized light (Fig. 15). The rate of crystal growth is dependent on the viscosity of the melt and can be hampered if the temperature is reduced very rapidly, ie. the melt viscosity is increased so rapidly that crystal growth and size is impeded. Indeed, very rapid cooling, such as when cold water is applied immediately after the mould is removed from the oven, can result in a freezing of crystal growth well below that which would normally have occurred had the melt been allowed to cool down at its own specific rate.

The degree of the resultant crystallinity will be reflected in the density of the solid material. In effect, a rapidly cooled part will have a lower crystallinity and hence a lower density, and conversely, a slowly cooled part will have greater crystallinity and hence higher density.

It is not always realized by the moulder how significant this can be. The cooling rate of a moulding is a vital part of the process as it can have a major effect on the end properties and behaviour of the product.

The grade of polyethylene supplied by the manufacturer will be given a nominal density that has been measured under controlled test conditions. The density of the rotational moulding, however, may be very different and dependent on the treatment it has been given during the cooling cycle. In other words the original density of the material can be significantly changed. This change is known as 'Density Shift'.

Table 7. Polyethylene Types: Nominal Properties

RESIN	DENSITY (g/cc)	MFI
A	0.940	3.7
B	0.935	4.0
C	0.925	5.0

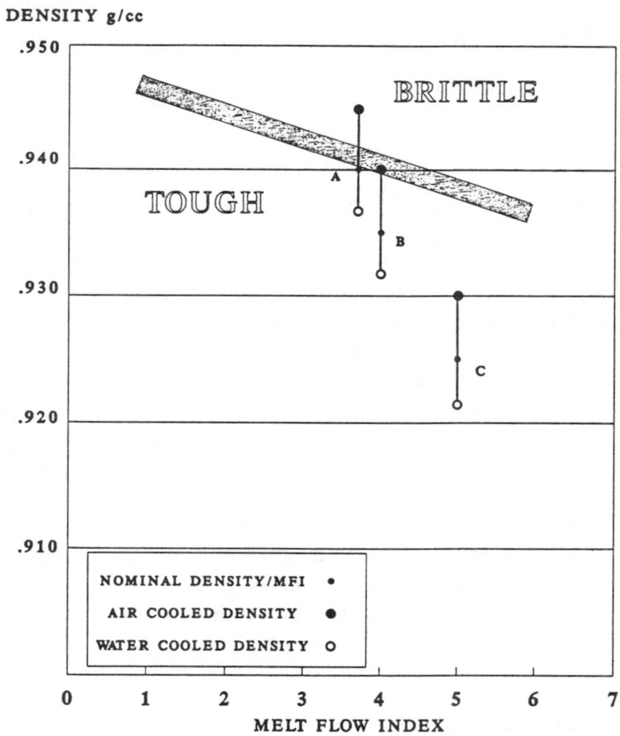

FIGURE 16
Effect of cooling on density (density shift)

The Effect of Density Shift

The effect of cooling on three different grades of linear polyethylene is shown in Fig.16. They each have nominal properties quoted on data sheets as shown in Table 7. These polymer resins are designed to produce tough impact resistant mouldings by ensuring that as the density is increased the MFI is decreased. It has been shown that the density can be raised or lowered according to the method of cooling employed in the rotational moulding machine. Rapid water cooling will reduce density, thus maintaining the tough impact resistant properties. On the other hand slow cooling in still air will raise the density to dangerously high levels and the resultant product can become impact brittle.

Slow cooling will result in products with a minimum of warpage or distortion. Rapid cooling will have the opposite effect and can cause severe distortion. The moulder is again faced with a compromise. The best shaped moulding, most closely resembling the ideal reflected by the mould itself, could have reduced impact properties.

A carefully programmed cooling cycle that employs air cooling, perhaps fan assisted, with a degree of controlled water cooling is usually sufficient to achieve normal crystallization and mouldings of the expected density. Should this not be the case it will be necessary to select a tougher grade of polyethylene, and the choice will then be between one of lower nominal density or of lower melt flow index.

6. APPLICATIONS

Since the earliest days in the development of the rotational moulding process many very successful products have been manufactured. The advances in the types of polyethylene available to moulders and a better understanding of the unique capabilities of the process, and its limitations, have been instrumental in the development of well designed products that in many cases could not have been economically produced by any other technique. Further advances have come with the development of flame retarded versions, semiconductive, antistatic and expandable or cellular formulations. Certainly the largest thermoplastic mouldings ever made in a single shot process are manufactured by rotational moulding, whilst a number of products are so geometrically complex that no other process could have been contemplated. A few examples of rotational moulded products are given in Table 8. However, there have been a number of dismal failures caused mainly by a lack of understanding of the polymer properties, the influence of the powder characteristics, inadequate knowledge of the process variables, and a lack of production control. It is hoped that this chapter will play a part in overcoming some of these problems.

REFERENCES

1. T.O.J.Kresser 'Polyethylene', Reinhold Publishing Corp. New York, 1957.
2. P.Y.Kelly 'A Microscopic Examination of Rotomoulded Polyethylene', Du Pont Canada Inc.
3. Information supplied by Additives Division of Ciba Geigy AG.

Table 8. Typical Rotomoulding Applications

MATERIALS HANDLING PRODUCTS

Tanks (Agricultural, Chemical, Fuel, Septic)
Chemical IBC's (Intermediate Bulk Containers)
Chemical drums, Shipping containers
Double walled (insulated) bins, Tote boxes
Wheeled bins, Hoppers, Coal bunkers

INDUSTRIAL PRODUCTS

Pump housings
Pipe fittings
Effluent ducts, Air ducts, Sewer linings
Vehicle consoles, Vehicle exterior trim
Safety Helmets, Stretchers
Light fittings

ENVIRONMENTAL PRODUCTS

Litter bins, Sanitation bins, Grit bins
Bottle banks

LEISURE PRODUCTS

Canoes, Kayaks, Windsurf boards, Boats, Trailers
Toys, Playground furniture, Domestic furniture
Point of display items, Mannequins, Planters

MARINE PRODUCTS

Floats, Buoys, Pontoons, Life belts

ROAD FURNITURE PRODUCTS

Road barriers, Road cones, Road signs

CHAPTER 3

Rotational Moulding of Nylons

F. Petruccelli

1. INTRODUCTION AND CHEMISTRY

Nylon or polyamide resins are long-chain polymers whose structures contain repeating amide groups (-CONH-). They were first introduced for fiber extrusion and their use has expanded into a wide variety of applications and processes. The range of properties possible is very broad and they can be processed by any of the major techniques used for thermoplastics. Chemically, nylons can be broken down into three primary classes: those nylons polymerized by the condensation of a dicarboxylic acid and a diamine:

$$HOOC\text{-}R\text{-}COOH + H_2N\text{-}R\text{-}NH_2,$$

those made by addition reaction of monomeric compounds that contain both acid and amine groups called lactams:

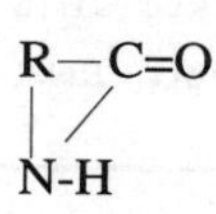

and those polymerized by self-condensation of amino acids:

$$H_2N\,(CH_2)_{10}\,COOH$$

An example of the first class of polyamides is nylon-6,6 which is polymerized from adipic acid and hexamethylene diamine:

$$-\!\!\left[CO(CH_2)_4\ CO\text{-}NH(CH_2)_6\ NH\right]\!\!-_n$$

nylon-6,6

The second class of polyamides is represented by nylon-6 which is formed by the addition reaction of epsilon-caprolactam:

epsilon-caprolactam nylon-6

The third class can be represented by nylon-11 which is made by the self-condensation of 11-aminoundecanoic acid:

$$H_2N\,(CH_2)_{10}COOH \longrightarrow \left[NH(CH_2)_{10}\,CO \right]_n$$

11- aminoundecanoic nylon-11

1.1 Nylon Numbering System

The nylon numbering system is based on the number of carbon atoms in each of the reactants used to prepare the particular nylon. The first number in a generic designation refers to the number of carbon atoms in the chain of the diamine and the second refers to the number of carbon atoms in the dicarboxylic acid chain. Nylon-6,6, as described above, is an example of this type. Where there is only one number, it means that the nylon was formed from a single monomeric material. The number will be the number of carbon atoms in the linear chain of the recurring polymer units. Nylon-6 and nylon-11 are the prime examples of this type.

1.2 Nylons Suitable for Rotomolding and Modifications

The diversity in types of nylon resins results from a wide choice of starting materials, variation in degree of polymerization, and modification through reaction or combination with other substances [1]. For example, nylon-6 can be manufactured in a plasticized form by leaving some or all of the residual unreacted monomer, caprolactam, in the product instead of removing it through a washing process. Also, nylon-6 can be reacted with an ionic copolymer, polyethylene methacrylic acid neutralized with zinc ions, or similar materials, to produce a copolymer nylon resin. A nylon copolymer can provide good rotomolding process capability as well as good compatibility and adhesion to most grades of polyethylene in a multi-layer rotomolded product [2] which will be discussed later in this chapter. Nylon-11 and nylon-12 can be combined with long-chain diols or high molecular weight sulfonamides to obtain a plasticizing effect. This increases flexibility and impact strength with a reduction in tensile and flexural properties [3].

Although a great many nylon resin types have been prepared and studied, there are relatively few actually being commercially produced. In addition, not all nylons have been found suitable for the rotational molding process. Nylon-6,6 yields inferior falling weight impact strength compared to nylon-6 when exposed to the high temperature and extended cycle times demanded for this high melting point resin (225°C) in the rotomolding process. Nylon-4,6 with a high melting point of 285°C, has also been evaluated for rotational molding and has been found to lack the good moldability and falling weight impact strength usually demanded for hollow-shaped rotomolded parts. Nylon-6, 9, nylon-6, 10, and nylon-6, 12 have melting points of 205°C, 215°C, and 219°C which should make them good candidates for the rotomolding process; however, they are not widely used in the rotomolding industry. At the present time, the most commercially successful nylons for rotomolding are nylon-6, nylon-11, and nylon-12 with melting points of 215°C, 186°C and 178°C, respectively. A comparison of properties of some commercially available nylon resins is compiled in Table 1.

2. MANUFACTURE OF NYLON RESINS

2.1 Nylon-6

The basic raw materials for nylon-6 are crude oil and natural gas. Fig. 1 is a schematic showing conversion of benzene to nylon-6 resin. Benzene is obtained from crude oil processed in refineries, and propylene and ammonia are derived from either crude oil or natural gas. To produce caprolactam, benzene and propylene are first reacted to form cumene which is then converted to phenol. When phenol is combined with ammonia and hydrogen, caprolactam is the final product.

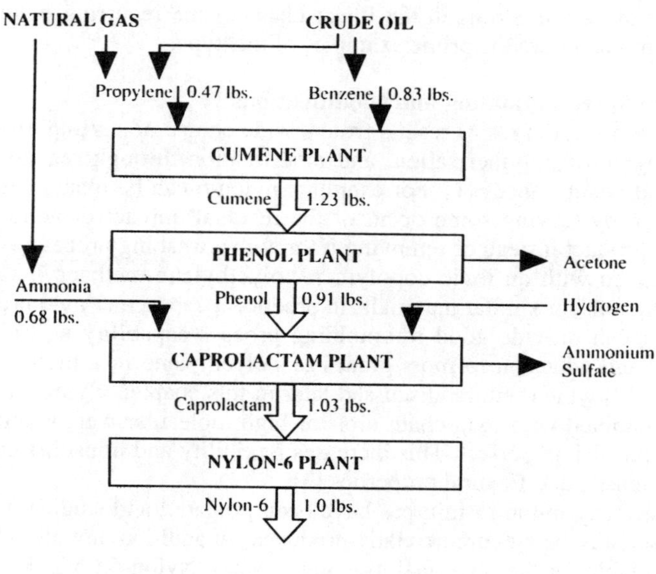

FIGURE 1 Benzene to nylon-6 conversion schematic

Table 1. COMPARISON OF PROPERTIES OF NYLONS, INJECTION MOLDED, DRY AS MOLDED

Properties	Units	ASTM Method	Nylon 6[a]	Nylon 66[b]	Nylon 11[c]	Nylon 12[d]	Nylon 6,9[e]	Nylon 6,10[f]	Nylon 6,12[f]
Specific gravity	-	D792	1.13	1.14	1.04	1.02	1.08	1.09	1.06
Melting point	°C	D789	215	255	186	175	205	215	219
Water absorption									
24-hour immersion at 23°C	%	D570	1.6	1.2	0.30	0.25	0.44	0.36	0.25
50% RH at 23°C	%		2.7	2.5	0.80	0.70	1.3	-	1.3
Saturation	%		9.5	8.5	1.9	1.5	4.5	-	3.0
Tensile strength at yield	MPa[g]	D638	82	83	57	50	70	70	61
Elongation at break	%	D638	150	60	330	200	50	-	150
Flexural modulus	MPa[g]	D790	2825	2827	980	1400	2310	2300	2030
Impact strength - Izod notched	J/m[h]	D256	65	53	40	50	37	45	53
Heat deflection temperature	°C	D648							
at 0.455 MPa (66psi)			185	235	150	150	168	173	180
at 1.85 MPa (264psi)			64	90	55	55	75	-	90
Continuous service temp. in hot air	°C	-	121	121	100	100	-	-	-
Rockwell hardness, R scale	-	D785	119	121	108	105	111	114	114
Coefficient of linear thermal expansion	mm/mm°C	D696	$8.3x10^{-5}$	$7x10^{-5}$	$9.2x10^{-5}$	$11x10^{-5}$	$15x10^{-5}$	$15x10^{-5}$	$9x10^{-5}$
Specific heat	-	-	0.4	0.4	0.5	-	0.4	0.4	0.4
Dielectric strength - short time	kV/mm	D149	17	15	17	18	22	19	14
Dielectric constant at 1000 cycles	-	D150	3.7	3.9	3.7	3.5	4.8	3.5	4.0

(a)Ref.4; (b)Ref.5; (c)Ref.6; (d)Ref.7; (e)Ref.8; (f)Ref.9; (g) To convert MPa to psi,multiply by 145; (h) To convert J/m to ft-lbs/in., multiply by 0.019.

- **Caprolactam Storage (Molten)**
  89°C
- **Mix Tank**
  Caprolactam + Additives + 4%H_2O
- **Autoclave**
  Pressure Cycle
  260°C, 55 PSIG
  Ring Opening, Addition
  Vacuum Cycle
  260°C, Start at 10 PSIG
  Polycondensation
- **Melt Extrusion**
  Strand and Granulate
  85 PSIG Nitrogen
- **Leaching**
  110°C Water, Extractibles <1-2%
- **Drying**
  Rotary Vacuum Dryer

FIGURE 2 Nylon-6 batch polymerization

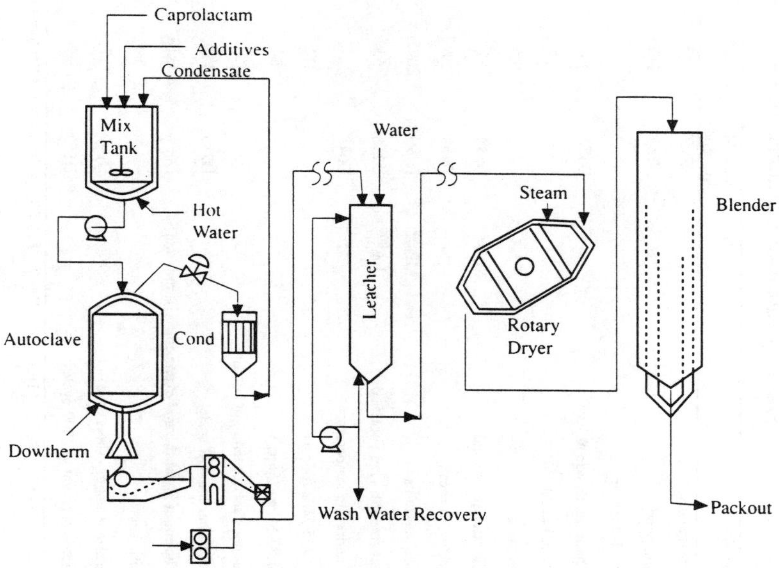

FIGURE 3 Nylon-6 batch polymer processing

Nylon-6 is manufactured by various procedures. One procedure is illustrated in Fig.2 and Fig. 3, showing caprolactam polymerization to nylon-6. In the polymerization of a rotomolding grade of nylon-6, it is preferred that more chain ends are terminated by acid groups than by amine groups [10]. This can be accomplished by the addition of a dicarboxylic acid, such as sebacic acid, during the polymerization process. The presence of more acid chain ends in the nylon-6 polymer provides the best resistance to oxidation during the heating cycle of the rotomolding process.

2.2 Nylon-11

Nylon-11 is not produced from coal or petroleum derivatives, but is manufactured from castor oil, a renewable, agricultural source. The alcoholysis of castor oil produces a methyl recinoleate which by pyrolysis generates undecylenic acid [11]. The 11-aminoundecanoic acid, which is the monomer for nylon-11, is made by the reaction of undecylenic acid with hydrogen bromide, followed by reaction with ammonia.

2.3 Nylon-12

Nylon-12 is polymerized from the monomer, lauryllactam, a petrochemical derivative. The basic starting material, butadiene, is converted by catalyzed reaction to the butadiene trimer and then to oxime and lactam product. Nylon-l2 was first produced in a commercial plant in Germany in 1966. The nylon-12 is polymerized from lauryllactam by processes similar to nylon-6; however, polymerization is slower because of the stability of the 12-carbon ring [12]. Nylon-12 is currently used more extensively than nylon-11 for rotational molding. It appears to provide more consistent falling weight impact properties than the nylon-11 in the rotomolding process.

3. COMPOUNDING

Nylon base resins for rotational molding require the addition of anti-oxidants and/or heat stabilizers before they can be sold as engineering resins. Nylons are prone to thermal degradation to varying degrees depending on composition, and therefore, must be stabilized with anti-oxidants or processed with an inert atmosphere inside the mold during the rotomolding process. The additives are normally compounded into nylon resins in a compounding unit shown schematically in Fig. 4.

3.1 Nylon-6

Some nylon-6 rotational molding resins can be molded without the use of an inert atmosphere if an inferior surface appearance and some reduction in impact properties can be tolerated by the rotomolder. In order for this to be possible, the nylon-6 composition should have chain ends substantially terminated with acid (carboxyl) groups which resist thermal oxidation much better than amine terminated or unterminated resins. In addition, nylon-6 resin should be unplasticized (i.e., washed to less than 2% residual caprolactam) Nylon-6 resins containing about 10% caprolactam as plasticizer will undergo severe oxidation during the rotomolding process

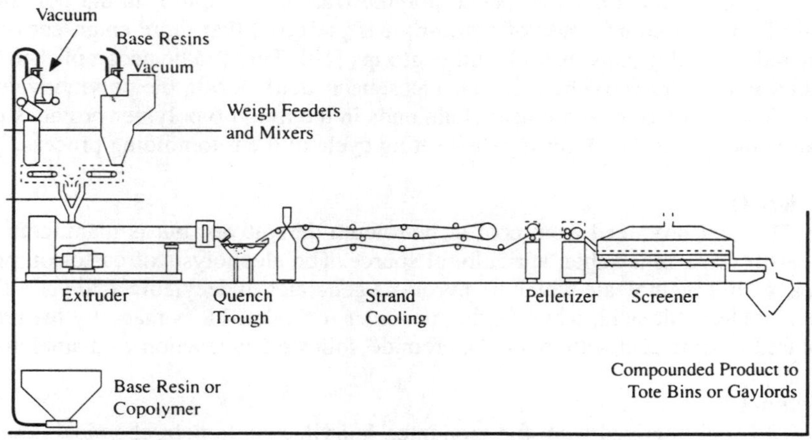

FIGURE 4 Typical nylon compounding unit

unless the mold is purged during the the heating cycle with an inert gas such as nitrogen or carbon dioxide.

Also compounded into nylon-6 rotational molding resin is a heat stabilizer system which provides heat resistance to the final molded product when exposed to elevated temperatures above 75°C (167°F). The heat stabilizer systems most widely used are usually based on copper salts, such as cuprous iodide combined with potassium iodide for longer term heat resistance. In addition to long term heat resistance, it is theorized that there is some interaction of the copper cation with the nylon-6 polymer which contributes to the reduction in the degree of crystallinity by virtue of disordering of the structure [13]. A disordered polymer structure will cause a lower degree of crystallinity and result in a higher falling weight impact strength in the final product. Fast cooling during the rotational molding process is also important in reducing the degree of crystallinity of the nylon-6 polymer for high impact strength.

3.2 Nylon-11 and Nylon-12

Rotomolding grade nylon-11 and nylon-12 resins are compounded with anti-oxidants and therefore do not require purging of mold with an inert atmosphere (nitrogen gas) during the heating portion of the rotomolding process. However, it has been found that better color retention and falling weight impact properties are achieved when these resins are rotational molded using an inert gas purge during the heating stage. Heat stabilizer systems are also compounded into nylon-11 and nylon-12 for good, long term heat resistance in the final molded product.

4. PROCESSING

4.1 Moisture Considerations

Rotomolding grade nylons are hygroscopic and will undergo hydrolytic degradation if processed with excessive moisture. As moisture absorption is directly dependent on surface area exposed, rotomolding nylon powders (usually 35 mesh or 20 mesh) more readily absorb moisture when exposed to the air than the 3.2 x 3.2mm (.125 x .125 in.) pellet product supplied for other molding processes. Nylon-6 rotomolding grade materials when supplied as smaller dry pellets (usually 0.8mm-1.6mm in diameter and 0.8mm-2.5mm long) can be rotomolded as received for many applications. For parts 3.2mm (.125 in.) thick or greater, nylon-6 does not have to be ground to a powder before molding. The pellets will readily flow out during the heating cycle to form a solid unitized structure. Where thinner wall thickness is required or where difficult to coat areas of a mold are encountered, nylon-6 ground to a 20 mesh or 35 mesh powder must be used and greater care exercized to prevent excessive moisture absorption before molding.

All nylon resins, both pellets and powder, are supplied dry in sealed containers ready for rotomolding. Once containers are opened, the nylon should be rotomolded and any unused resin resealed in heavy duty polyethylene bags (thickness-greater than 0.lmm or 0.004 in.) for storage to prevent moisture pickup.

It should be noted that nylons are not as sensitive to moisture as polycarbonate in the rotomolding process. For satisfactory performance in processing, nylons should not contain more that 0.2% moisture before rotomolding. In the case of polycarbonate resin, the moisture content must be less than 0.01% which usually requires that the rotomolder dry the resin immediately prior to addition to the mold. On the other hand, most nylons are supplied dry and will provide the rotomolder

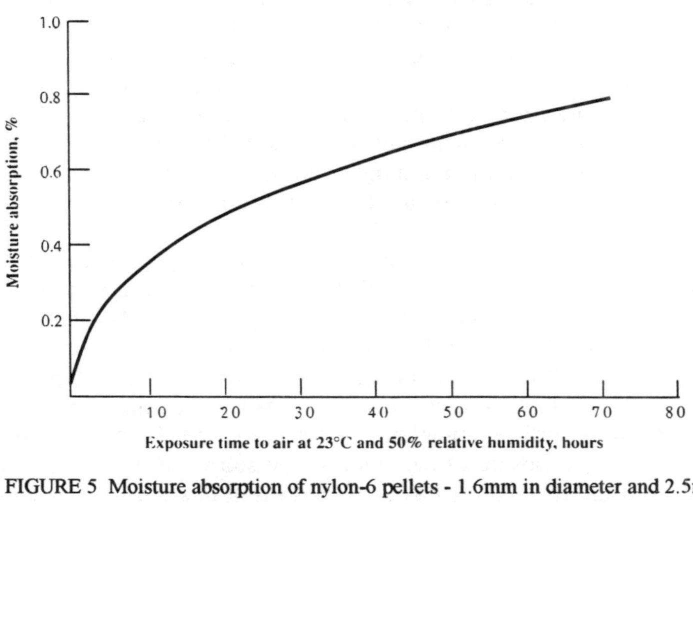

FIGURE 5 Moisture absorption of nylon-6 pellets - 1.6mm in diameter and 2.5mm long

with sufficient latitude in moisture control before molding without need to dry. Nylon-6 pellets supplied in 25kg (55 lb.) moisture resistant bags can normally be rotomolded without undue concern over excessive moisture pickup prior to molding, if exposure time is limited to an 8 hour shift. Fig.5 plots moisture absorption of nylon-6 pellets vs. time at 23°C and 50% R.H. In the case of 35 mesh nylon-6 powder with a much greater surface area, more care must be exercised to minimize moisture pickup from the atmosphere prior to molding. One way to avoid moisture pickup in nylon powders is to weigh out the exact mold charges in polyethylene bags and to seal bags with a rubber band or plastic tie until ready for molding. A good practice, after filling the polyethylene bag with nylon, is to twist the top of the bag and bend it over to make it air tight before securing it with a rubber band.

Nylons will absorb moisture at different rates and to different levels as shown in Table 2.

TABLE 2. ABSORPTION OF MOISTURE BY NYLONS (% BY WEIGHT-DRY BASIS)[a]

Material	24hr. immersion ASTM D-570 (3.2mm or 0.125 in.)	Equilibrium at 50% R.H.	Saturation water immersion
Nylon-6	1.60	2.7	9.5
Nylon-11	0.30	0.8	1.9
Nylon-12	0.25	0.7	1.5

a.Data from trade literature.

As can be seen in Table 2, Nylon-11 and Nylon-12 absorb much less moisture than nylon-6. However, the lower sensitivity to moisture of nylon-11 and nylon-12 compared to nylon-6 is mainly significant in the final rotomolded part, where it provides better dimensional stability and % retention of mechanical properties when exposed to various moisture conditions. Prior to processing, nylon-11 and nylon-12 resin manufacturers recommend not to exceed 0.1% moisture content in the rotomolding grade resins for optimum molded properties. In the case of nylon-6, resin suppliers recommend materials not exceed 0.2% moisture prior to rotomolding.

If pellet or powdered nylon resin containers are not properly sealed, or if the resin is exposed to moisture from the air such that it exceeds the recommended moisture level, drying may be necessary for satisfactory processing. Drying is best accomplished using vacuum dryers, but dehumidifying hopper and oven dryers can be used. The recommended drying temperature is 80°C (176°F) at a dew point of -23 to -35°C (-10 to -30°F). Drying in air above 80°C is not recommended due to the possibility of excessive surface oxidation of resin particles. The drying time is dependent on the extent of exposure of the nylon to moisture and the particle size of the resin. Short exposure times of nylons to air will result in surface moisture on the resin particle which can readily be dried, usually within a few hours. Extended exposure times will result in moisture absorption deep within the particle, and require as much as 24 hrs. for satisfactory drying of the nylon. Fig. 6 illustrates the drying rates for nylon-6 homopolymer resin which has been exposed to two different levels of moisture, 1.0% and 0.45%.

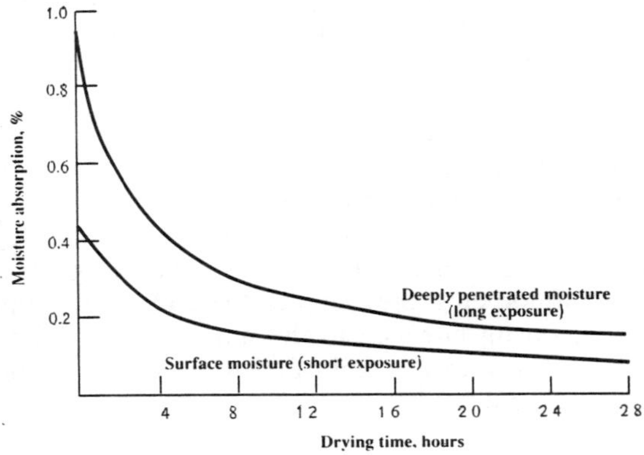

FIGURE 6 Drying of nylon-6 at 80°C

4.2 Machine Considerations

4.2.1 Machine Type

Although nylons specifically designed for rotational molding can be processed in most types of rotomolding machines, the most suitable are the machines which permit continuous, uninterrupted rotation of the arm from the heating to the cooling chamber. This continuous rotation will prevent 'sagging' (running) of resin melt, especially when molding thick parts. Rotational molding grade nylon resins in the melt stage can become very fluid at the completion of the heating cycle, and an interruption or change in rotation of the arm of the machine could cause 'sagging' of resin along the mold surface leaving sections of the part wall with considerable variation in thickness.

4.2.2 Inert Gas Injection

To prevent discoloration and loss of some molded properties due to oxidation, nylon-6 resins require the use of an inert atmosphere within the mold. This is most conveniently accomplished by introducing nitrogen or carbon dioxide gas into the mold during the rotomolding heating cycle.

Most rotomolding machines have arms which are drilled for gas injection, so that it is only necessary to connect a gas cylinder to the arm through rotating valves. Only a small flow of nitrogen into the mold is required to provide a satisfactory inert atmosphere. The larger the mold, the greater the flow of nitrogen that will be required to displace the air in the mold cavity. Adequate purging is evidenced by a light-colored interior of molded part. It should be noted that a good seal is required around the head assembly of the rotating arms of the machine to ensure that nitrogen does not leak excessively, causing insufficient gas to reach the mold and waste of nitrogen gas. Routine maintenance checks should be made of 'O' ring seals. An audible air leak around head assembly or grease contamination inside the molded part are signs of leaking 'O' ring seals.

4.2.3 Methods of Rotomolding using Nitrogen

There are two methods of rotomolding with nitrogen purging:

(1) an 'open system' where gas is injected into a vented mold, and

(2) a 'closed system' where gas is injected into a non-vented mold. In the 'open system', nitrogen is introduced in the hole in the mold by connecting a length of PTFE (Teflon®,[a] polytetrafluoroethylene) tube to the end of the copper tubing (use hose clamp) delivering nitrogen to the mold. The PTFE tube section of nitrogen line is placed into the mold through a larger diameter standard PTFE vent tube as illustrated in Fig.7.

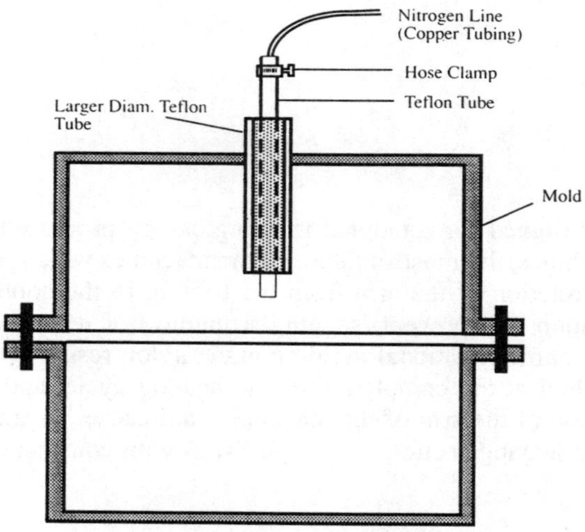

FIGURE 7 Open system: nitrogen into single vent hole

Another 'open system' method is shown in Fig.8, where nitrogen is fed into the mold at one opening and a standard PTFE vent tube installed in another opening in the mold. The diameter of the standard vent will determine the pressure developed

(a) Registered trade mark of E.I. duPont de Nemours & Co.

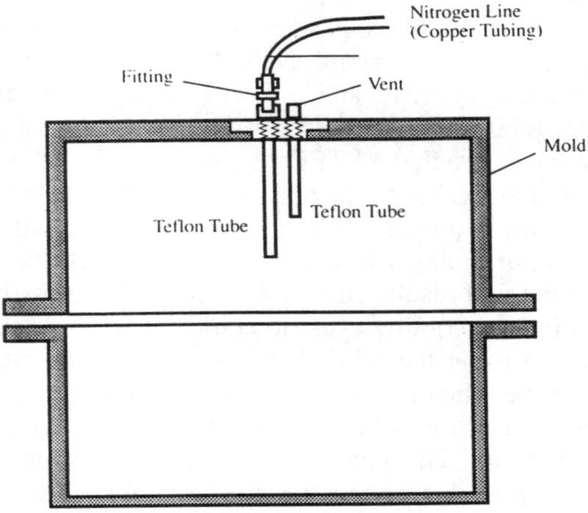

FIGURE 8 Open system: nitrogen into vented mold

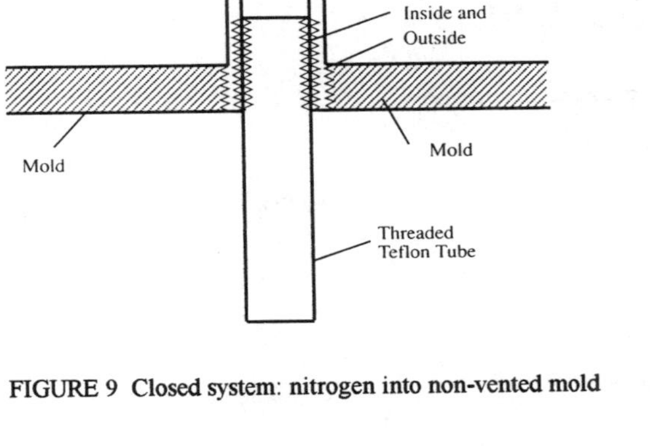

FIGURE 9 Closed system: nitrogen into non-vented mold

inside the mold after the parting line is completely sealed with melted resin during the heating cycle. Use of pressure during the cooling cycle will control warp and provide more efficient cooling of part (see cooling cycle 4.4.3). The smaller the vent tube diameter, the greater the pressure that can be developed within the mold.

In the 'closed system', copper tubing is connected directly to the mold as illustrated in Fig.9. It can be seen in the figure that a PTFE tube is inserted well into the inside threads of a pipe section which has been attached to the mold. It is important to have only PTFE tubing protruding into the mold since very little resin will build up on the PTFE. Any metal tubing or piping will conduct so much heat that undesirable build-up of resin on the tube will result. The 'closed system' permits higher levels of pressure to be used during the cooling cycle to control warp and improve cooling efficiency. It is essential in using the 'closed system' to stop the nitrogen flow at the point in the heating cycle when melted resin has completely sealed the parting line of the mold, so that no resin will be extruded into the parting line by excessive pressure build-up in the mold. This can be accomplished by shutting the nitrogen source off and by venting back through the arm assembly. Fig.10 is a schematic of a complete nitrogen purging hook-up showing a valve for venting through the drilled out spindle of the machine arm. Also illustrated in Fig.10 is a pressure relief valve which is essential for use with any gas injection method as a safety precaution in the event of excessive pressure build-up within the mold.

If the arm is not drilled out for gas injection, carbon dioxide in the form of dry ice cubes (pieces about 12 cu.cms. or 0.75 cu.in.) can be added to the mold cavity

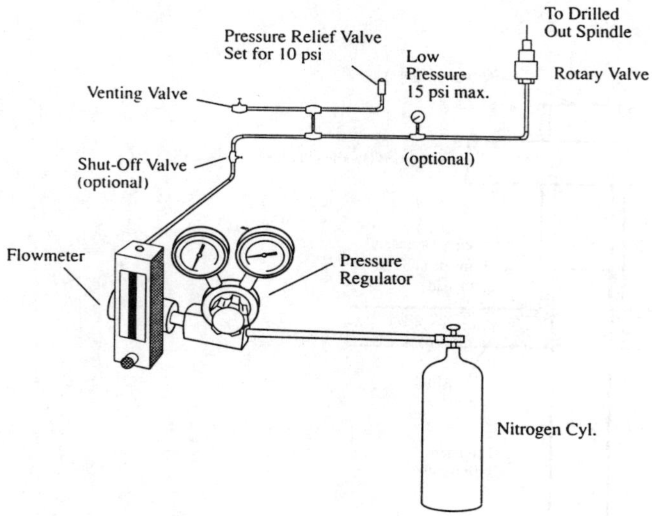

FIGURE 10 Nitrogen purging hook-up

at the same time the resin is charged to mold. The pieces of dry ice will slowly sublime during the heating cycle and provide an inert atmosphere throughout the entire molding cycle. Generally 7-13 gm. per liter (25-50 gm./gal.) of mold volume is sufficient to yield a satisfactory inert atmosphere. Besides the inconvenience of its use, dry ice can cause poor surface appearance in molded part.

4.3 Molds and Mold Release Coatings

4.3.1 Molds

Nylon resins can be rotomolded in molds made from standard materials: cast aluminum, sheet aluminum, sheet steel, stainless steel, and electroformed nickel. Because nylons flow better than polyethylene resins, tools must be built more carefully to avoid excessive resin flow into the parting line of the mold. Cast aluminum molds are the most widely used type of mold for nylon resins. The excellent heat transfer characteristics of cast aluminum are especially beneficial for the high melting point nylon resins in order to achieve fast molding cycles. Cast aluminum molds also permit the use of tongue and groove parting lines which provide better seating of mold and less tendency for resin flashing into the parting line. Fabricated stainless steel molds are also well suited for use with nylons; however, they can be difficult to produce for very complex shapes. If nylon part dimensions are critical, molds are best fabricated with 10 gauge (3.4mm) stainless steel. A 14 gauge (1.9mm) stainless steel mold fabricated into a complex shape is more apt to result in excessive distortion when rotomolding at the high oven temperatures used for nylon.

Nylons can be rotomolded with a smooth or mat finish, exhibiting the finish found on the mold surface. However, a smooth surface is preferred since any mold release coating baked-on surface can more uniformly be sanded if required, to control release of a part from the mold surface during the cooling operation. Because nylons are much stiffer than polyethylenes, to facilitate part ejection it is good mold design practice to: (1) avoid any undercuts, (2) use approximately a 1.5° draft angle, and (3) use radii greater than 9.5mm (0.375 in.).

4.3.2 Mold Release Coatings

Nylons release more easily from the mold surface than polyethylene and care must be taken that applications of mold release coatings are not excessive in order to avoid premature release and consequent warping, surface pitting, excessive shrinkage, and low impact strength of molded parts. Most baked-on type mold release coatings (for example Frekote 44[(a)] or Chemtrend C24[(b)]) can be successfully used to provide satisfactory release of molded nylons from the mold surface.

The mold release coating should be applied thoroughly, but sparingly, to the mold surface and then baked in an oven according to manufacturer's instructions. After baking, the mold surface usually must be sanded using Scotchbrite[(c)] or 80-

(a) Registered trademark of Hysol, The Dexter Corp.,Seabrook, N.H.
(b) Registered trademark of Chem-Trend, Inc. Howell, MI.
(c) Registered trademark of 3M Co., St. Paul, MN.

200 grit emery cloth to prevent premature release and maintain good contact with the mold surface during the cooling cycle. (When internal pressurization of the mold is used during the cooling cycle, sanding of mold release coating on the mold may not always be necessary since the pressure will hold the solidified polymer against the mold walls and prevent premature release from molds.) When premature release occurs, less than optimum impact strength will result due to the development of a high degree of crystallinity caused by the resulting slow cooling rate.

Nylons can be rotomolded with very little mold release coating on a highly polished electroformed nickel mold. In this case a high luster, smooth surfaced part can be obtained with a 20 mesh nylon powder.

4.4 Molding Conditions

With a properly designed mold, a means of achieving an inert atmosphere within the mold, and a satisfactorily prepared mold surface, nylons can readily be rotomolded whether in pellet form or as a ground powder. Slow spindle rotation speeds have been found more suitable for pellets since they have more difficulty clinging to the mold surface, especially on sharp, deep-pocket areas in a mold. The pellets, although having ideal particle shape and pouring characteristics, are inferior to powder in surface appearance and uniformity of wall thickness, and require longer cycle times. Nylon-11 and nylon-12 rotomolding resins are supplied only as 35 mesh powders. Nylon-6 is commercially available in either pellet or powder form.

In some applications, rotomolders charge the mold with a mixture of nylon-6 pellets and powder to achieve the molding performance desired. During the heating cycle the powder will deposit on the mold first, followed by the pellets to provide improved moldability compared to an all-pellet resin charge.

As with polyethylene resins, for best results, nylon powders must be properly ground within certain limits of coarse and fine fractions and with a minimum of tails or tentacles protruding from the particle. Powders that contain an excess of particles with 'tails' will exhibit poor powder dry flow in the mold and cause 'bridging' in some mold designs. A good nylon powder grind is best produced cryogenically to obtain a low amount of 'tails' with the added benefit of higher production rates.

The rotation ratio used with nylons is dependent on the configuration of the part to be molded and is generally identical to the rotation ratio that would be used for polyethylenes.

4.4.1 Charging the Mold

Most nylon pellets and powders are supplied dry, ready for rotational molding. The resin is taken from an unopened or sealed container (air tight to prevent moisture pick-up) weighed out to required charge weight and immediately placed into the mold. Some rotomolders prefer to pre-weigh the exact charge weight of nylon into polyethylene bags prior to molding. In this case care must be taken to properly seal the polyethylene bag to prevent moisture pick-up.

If nitrogen purging of the mold is to be used, the nitrogen flow rate should be established (rotameter reading) by opening the nitrogen line control valve sufficiently to obtain a small but perceptible stream of nitrogen at the point of entry into the mold. The PTFE tubing conveying nitrogen inside the mold should not be placed too close to a mold surface so as not to have the nitrogen blowing onto the mold surface and preventing resin build-up in the area. Adequate flow of nitrogen during the heating cycle will produce a non-oxidized, light-colored interior of the molded part.

The mold is filled with the required amount of resin, mold vents are installed, the mold is clamped shut, and a nitrogen line is attached to the mold. The mold is indexed into a high temperature oven, and nitrogen flow into the mold is initiated. A sufficient nitrogen flow is required to effect purging of air out of the mold through the parting line and to prevent hot, high velocity oven air from entering the mold and disrupting the inert atmosphere inside the mold.

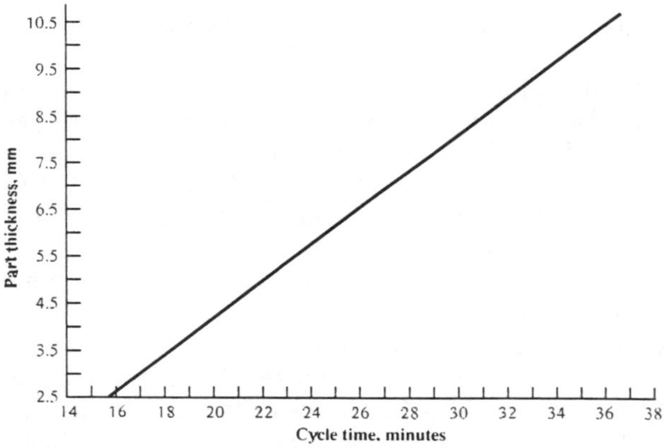

FIGURE 11 Cycle times vs. thickness - nylon-6 pellets

4.4.2 Heating Cycle

Oven temperatures of 260 to 360°C (500 to 680°F) are generally employed for nylon resins. Higher processing temperatures could be used with nylons for increased production rates, but are usually not the choice in order to avoid the possibility of higher mold and machine maintenance costs. Since nylon-12 has a lower melting point than nylon-6 (178°C vs. 215°C), it generally is processed at lower oven temperatures than nylon-6. Fig.11 shows cycle time versus wall thickness for nylon-6 pellets rotomolded in a mold constructed of 6.35mm (0.25in.) aluminum sheet. The oven temperature used was 343°C (650°F). A nominal 3.2mm (0.125in.) thick part weighing 2.27kg (5lbs.) under the above conditions required a minimum of 18 minutes to achieve proper fusion of the pellets in a laboratory machine (McNeil 500-48). As a rule-of-thumb, two minutes additional or less time is

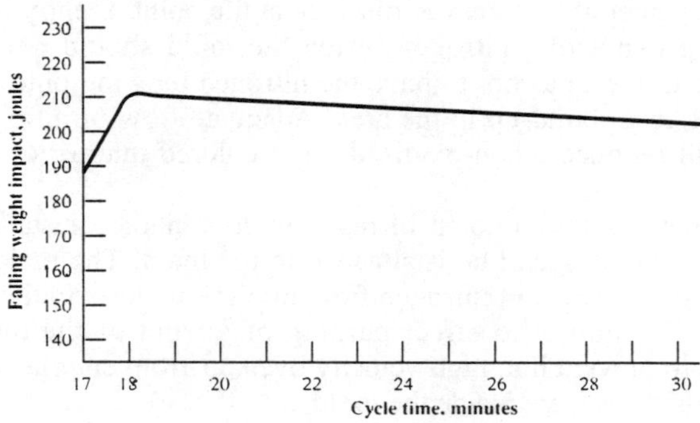

FIGURE 12 Falling weight impact strength (23°C) vs. cycle time, nylon-6 pellets

required per 0.76mm (.030in.) wall thickness added or substracted in the above molding system. Larger machines with more efficient heating systems will permit shorter heating cycles. With a 35 mesh grind, a nylon-6 rotomolding resin powder would achieve proper fusion at a minimum cycle of about 14 minutes at 343°C for a nominal part thickness of 3.2mm (0.125in.). For the same part, the minimum cycle time for nylon-12 would be about 12 minutes.

The point of minimum good fusion of nylon rotomolding resins is determined when all particles (pellets or powder) are completely flowed-out, providing a smooth surface on the inside surface of the molded part. Any particles which are imbedded into inside surface of the molded part and which are not completely flowed-out will seriously affect the molded properties of the part, especially its falling weight impact strength. It is good commercial practice to add 3-4 minutes to the minimum fusion time established for a nylon molded part to ensure well cured parts. In Fig. 12 the falling weight impact strength of a rotomolding grade of nylon-6 (pellets) tested at 23°C is plotted versus heating cycle time. As can be seen, once good fusion of the pellets is achieved (in this case at about 19 minutes), there is minimal loss in impact strength even after extending the cycle by 12 minutes. The use of nitrogen purging prevents oxidation and thereby permits substantial leeway in the extension of the heating cycle, if desired.

If nitrogen purging of the mold is used, the flow of nitrogen gas can be stopped at a point in the heating cycle when melted resin has covered the mold parting line completely, preventing any high velocity oven air from entering the mold. The proper time to cut off nitrogen can be easily determined by trial-and-error. For the nylon-6 pellets used for Fig. 11, the nitrogen cut-off time would be about 10-13 minutes. Where a vented mold is used, the nitrogen flow need not be stopped, but can proceed right through the complete molding cycle. In a mold with only a nitrogen inlet and no vent, once parting is sealed with resin, it is essential to stop the flow of nitrogen to prevent pressure build-up which would cause unwanted resin flow into the parting line. In the latter case the mold should be vented back through the arm assembly (see Fig.10).

4.4.3 Cooling Cycle

Cooling the mold is especially important for nylon resins because they are semi-crystalline materials. Understanding the benefits of improved cooling techniques can reap great benefits in impact properties as well as in reduced shrinkage and warpage. Once the nylon resins have been fused into a highly viscous layer on the inside mold surface, the mold is indexed into the cooling chamber of the rotomolding machine where it is first cooled by air and then by water spray or mist while still being biaxially rotated. It is best if the rotation ratio remains the same as that used in the oven cycle in order to avoid the possibility of melted resin movement along the internal surface of the mold. This possibility is greatly increased in thick-walled parts (greater than 6. 3mm).

In the molten state, nylon resins are amorphous (non-crystalline) with molecules moving independently and completely disordered. During solidification, the molecules stop moving and some of them will arrange themselves into a regular pattern of fixed positions. This phenomenon is called crystallization. This process of forming crystals or spherulites continues until the temperature falls below the glass transition temperature, Tg, of the particular nylon. Below the Tg, the molecular motion ceases and very little additional crystalline development occurs. The longer it takes to quench the nylon from the melt temperature to below the Tg, the higher will be the percent crystallinity and the lower will be its toughness and impact resistance. In other words, the rate of cooling of semicrystalline nylons directly affects the number and size of spherulites which in turn affect the molded properties of the polymer.

In rotational molding, unlike other processes, quick quenching of the polymer melt requires greater care in order to produce satisfactory parts. Once the mold reaches the cooling chamber, if water is immediately sprayed onto the hot mold, premature release of nylon part from mold will occur and result not only in inefficient cooling (due to loss of resin contact with mold surface) but also in severe warpage. Once premature release from mold surface occurs, an air insulating gap between mold surface and part will develop which will cause poor transfer of heat from part through mold and result in undesirable slow cooling of molded part. For this reason a cooling cycle for nylon should consist of a short period of air cooling followed by an adequate fine water spray. An optimum cooling cycle for semi-crystalline nylon resins should utilize the minimum air cooling time that serves to prevent premature release, followed by a period of water cooling. Part design often dictates how much to cool and when to remove the part from the mold, but in general for a 3.2mm (0.125in.) thick part, 5 minutes of air cooling followed by 5-10 minutes of water spray should provide an adequate cooling cycle. A typical molding cycle for nylon-6 pellets is listed in Table 3.

TABLE 3. TYPICAL MOLDING CONDITIONS FOR 67 LITER NYLON-6 TANK

Part Thickness, mm	3.8 (.150 in.)
Approximate Part Dimensions, cm	40.6 x 40.6 x 40.6
Part Weight, kg	4.54 (10 lbs.)
Oven Temperature, °C	343 (650°F)
Heating Cycle, Minutes	22
Cooling Cycle, Minutes	
Air	5
Water	7
Air	2
Rotation-Major Axis, rpm	6
Rotation Ratio	4:1
Inert Atmosphere	YES

If slight gas pressure, 0.14 to 0.21 kgs/cm^2 (2 to 3 psig) can be introduced inside the mold during the cooling cycle, a 1 to 1.5 minute air cooling time can be used followed immediately with a water spray. The gas pressure should be introduced about 15-30 sec. before the water spray is activated to prevent any extrusion of resin melt into the parting line of the mold. The air cooling time, although short, will allow a solid resin skin to form at the parting line of the mold and prevent resin flow into the parting line. This process provides for a very fast and efficient cooling cycle. The more efficient the cooling of nylon, the lower the crystallinity and the higher the impact strength.

The use of mold pressurization during cooling also permits greater control over part shrinkage and warpage. Shrinkage of nylon in the rotomolding process, as with other semi-crystalline resins, depends on the mold surface treatment (mold release), the shape of the part and how it is constrained in the mold, the rate of cooling, and the degree of pressurization applied inside the mold. For example, in rotomolding nylon-6 with just 0.21 kgs/cm (3 psig) gas pressure during the cooling cycle, the typical shrinkage obtained would be about 1.2-1.5% compared to 1.8-2.5% for an unpressurized, unrestrained molded part.

If the part must be extracted from the mold on the 'hot' side to better facilitate removal due to the complexity or design of the part, subsequent rapid quenching can be achieved by immersing the part in tap water or filling it with water. If convenient, filling the part with water is preferred because it will create a more amorphous (non-crystalline) interior surface and thereby produce a part with significantly higher falling weight impact strength. In addition, internal water cooling will permit accelerated moisture absorption on the inside of the part which will also impart greater toughness and impact strength to the molded part.

Table 4 illustrates the effect of fast cooling and soaking of the inside surface of a

9.5mm (0.375in.) thick nylon-6 part on impact strength of part. It can be seen that quenching of the inside surface of the part increases impact by 37% and an additional 2 hr. soak results in a 63% increase compared to the dry-as-molded impact result.

TABLE 4. EFFECT OF INTERNAL WATER COOLING AND ABSORPTION ON IMPACT OF NYLON-6 PART[a]

Conditioning:	Falling Weight Impact Str. [Joules[b])
Dry as Molded	75[c]
Quick Quench	103[c]
2 Hour Soak	122[c]
45 Hour Soak	170[c]

a. Part 9.5 mm (0.375 in.) thick
b. To convert joules to ft.-lbs., divide by 1.36
c. Test specimens conditioned 7 days, 50% RH and 23°C prior to testing

4.5 Multi-component Rotational Molding

4.5.1 Processing Considerations

The rotomolding process and nylon resins are well suited for multi-layer rotational molded articles. Through choice of resin and layer position a number of constructions can be achieved which will optimize each resin property as well as minimize cost. For example, nylon-6 can be used on the exterior of a rotomolded part to provide strength and rigidity with an inner layer of polyethylene or fluoropolymer providing a surface acceptable for food contact use, acid resistance, or hot water resistance. In the event that the two materials desired are incompatible with one another, it is also possible to introduce an intermediate polymeric material which is compatible with both materials to serve as a binder.

The multilayer rotomolding process can be accomplished in a number of ways depending on the particular design of the part. The formation of a multi-layered part by a sequential process is straightforward, with the first resin charged into the mold, heated until it is completely fused, and then the second charge is introduced and fused. There are several significant problems associated with this type of process, namely:

(1) Bonding compatibility of the resins
(2) Shrinkage differential of the resins
(3) Viscosity and thickness of first resin melt, such that flow (sagging) will occur when rotation is stopped during charging of second resin.

Although multi-layered rotomolded parts can be produced with nylons mechanically bonded to another resin layer, such as with nylon-6 and high density polyethylene resin (HDPE), the better co-rotomolded product would be one which develops some chemical bonding between layers. The chemical bonding can be effected either by modification of the nylon resin, by use of a modified resin for the other layer, or by using an adhesive ply chemically reactive with both the nylon and the other resin. For example, if the objective was to produce a good multi-layered product consisting of nylon-6 and high density polyethylene, this could be accomplished by using a nylon-6 graft copolymer made with a modified olefinic monomer and rotomolding it with a high density polyethylene resin. The polyethylene has good bonding capability with the olefinic modified nylon-6 copolymer. This construction could also be achieved with a homopolymer nylon-6 resin by utilizing a random copolymer olefinic product (such as ethylene acrylic acid copolymer resin) as an adhesive ply between the nylon-6 and HDPE.

4.5.2 The Two-step Multi-layer Rotomolding Process

The two-step process involves first charging the mold with one material which is desired to form the exterior of the part. After the first material is completely flowed out, the rotation of the machine is stopped and the second material is introduced into the mold through an access port in the mold. When nylon-6 is the initially charged resin, and the wall thickness of part exceeds 4.8mm (0.187in.), some 'sagging' of the resin melt may occur when rotation is stopped after the heating cycle. To avoid 'sagging', the second resin should be added only after a period of air cooling (5 to 15 minutes depending on the thickness of the outside layer) has taken place and the nylon-6 reaches a very viscous or semi-solid state. Sufficient heat may still be available in the nylon layer to melt the second resin within the air-cooling cycle.

There is an upper limit of the ratio of thickness of polyethylene to nylon-6 for which this technique will work. Theoretically the ratio is:

$$\frac{t\ PE^{a}}{t\ nylon\text{-}6^{\ b}} = 0.0065 \quad \left[\ (T_m - 400)\ .6 + 40\ \right]$$

a. Thickness of polyethylene layer
b. Thickness of nylon-6 layer

where T_m is the bulk temperature in °F of the nylon-6 resin upon adding the polyethylene inner layer [14]. For example, if T_m is 390°F and the nylon layer thickness is 0.250in., the maximum polyethylene thickness that could be obtained would be 0.055 in. If a thicker polyethylene layer were required, it would be necessary to return the mold to the heating chamber to effect complete flow-out of the resin. After the second material is fully laid down, the final air and water cooling of the part is completed in the cooling chamber. An example of a two-step multi-layer molding cycle is listed in Table 5.

TABLE 5. TWO-STEP MULTI-LAYER MOLDING CYCLE

Resin Outer Layer: Nylon-6 (pellets)
Resin Inner Layer: Linear Low Density Polyethylene (35 mesh)
Mold: 25.4 cm (10in.) Diam. Cast Aluminum
Machine: McNeil 500-48
Molding Conditions

Step #1:	Nylon charge weight	2.27kg (5lbs)
	Oven Temperature	343°C (650°F)
	Oven Time, Nylon-6 Resin	30 min.
	Nitrogen Purge, Open System	With Vent
		Duration 20 min.
	Rotation Speed, Major	8 RPM
	Rotation Ratio	4:1
	Nylon Resin Cooling Time	15 min.
Step #2:	Polyethylene Charge Weight	0.23kg (0.5 lbs)
	Rotation Time In Cooling Chamber	4 min.
	Final Cooling Time - Air	5 min.
	- Water	10 min.
Final Part Thickness	- Outside Layer	6.4mm (0.25 in.)
	- Inside Layer	1.6mm (0.06 in.)

4.5.3 The One-step Multi-layer Rotomolding Process

The one-step process of forming multi-layered parts involves the use of molds constructed with insulated internal hoppers. Here the exterior material is placed in the mold and the inner layer material placed in the internal hopper equipped with a pneumatically actuated door or cover plate. In this method the machine rotation is never stopped. Once the exterior resin has completely flowed out, the hopper door is actuated and held open for a short period of time by air pressure, manually or automatically by integration into the machine molding cycle. Once the contents have been completely dumped, the air is shut off and the hopper door closed. If a three layer construction is desired a second internal hopper arrangement would be required. This method is the most convenient and provides the shortest cycles; however, it has the limitation of requiring a fairly large size opening in the part in which to locate the insulated hopper.

4.5.4 Multi-layer Material Combinations with Nylon

There are several useful combinations of nylon with other resins which can be used in the multi-layer rotomolding process:

1. Polyethylenes - HDPE, LLDPE, Ethylene acrylic acid copolymer
2. Fluoropolymers: ethylene chlorotrifluoroethylene

It should be pointed out here that, although most polyethylene types can easily be used in combination with nylon in the rotomolding process, cross-linked polyethylene is an exception as an inner liner of a rotomolded part. Cross-linked polyethylene (CLPE) should not be added to a very hot layer of nylon melt because it contains a peroxide catalyst that will decompose violently when suddenly exposed to a temperature of 190-215°C, which can be encountered with melted nylon. This violent reaction will not produce an acceptable interior layer. The nylon outer layer should first be cooled sufficiently to avoid activating the catalyst at too high a temperature. Using CLPE as the outer layer does not present the same problem as described above, but will require an additional oven cycle, in order to completely fuse the higher melting nylon powder.

4.6 Nylon Reaction Injection Molding (RIM) adapted to Rotomolding

Although the process of anionic polymerization of caprolactam to produce nylon-6 resin has been known for many years, nylon RIM processing was not generally recognized until the mid-seventies with development work conducted by Monsanto Co., DSM (Netherlands), and Allied Corp. Nylon RIM is a block polymer which is produced by reacting anhydrous molten caprolactam with polymeric polyols, using a metal lactam catalyst. The nylon RIM processes developed all involve conditioning of the reactants, stoichiometric metering of two or more components, mixing, and then delivering mixed reactants into a heated mold designed to control the polymerization process.

In essence the RIM process consists of mixing anhydrous caprolactam with a metal lactam catalyst (e.g. caprolactam magnesium bromide), followed by degassing solution and then keeping the mix under a dry nitrogen blanket ready for mixing with another stream consisting of a prepolymer dissolved in caprolactam. The two reactive streams are then combined by impingement mixing or by use of a static mixer, and delivered into a mold. The separate streams must be maintained at an elevated temperature of 80-85°C (176-185°F) with the polymerization taking place within the mold maintained at 130-140°C (266-284°F) [15].

Anionic polymerization of nylon-6 into hollow shapes has long been known and practiced, but recent advances in RIM technology in the area of composition, reaction rates, and polymerization control have resulted in products more suitable for rotational molding. In 1990, DSM RIM Nylon Inc. introduced Nyrim Pl-30R (registered trademark of DSM RIM Nylon Inc. of the Netherlands and Georgia) prepolymer specifically designed for use in commercially available rotomolding equipment with some modifications [16]. The Nyrim P1-30R prepolymer, an activated polyether based polyol, is reacted with caprolactam to produce a nylon block copolymer of varying properties depending on the amount of prepolymer used. The higher the prepolymer content, the greater the toughness and the lower the flexural modulus. Table 6 provides some typical properties of Nyrim Pl-30 materials.

4.6.1 Rotomolding Processing of Nyrim Materials

As in the Nylon RIM process, the nylon block copolymer rotomolding process

TABLE 6. TYPICAL PROPERTIES NYRIM P1-30 MATERIALS[a]

Properties	Method	Unit	Nyrim 1000		Nyrim 1500		Nyrim 2000	
			Dry	Cond.[b]	Dry	Cond.[b]	Dry	Cond.[b]
Specific Gravity	DIN 53479	--	1.10	--	1.12	--	1.12	--
Melting Point	DSC	°C	214	--	214	--	213	--
Water Absorption 23°C, 50% RH	DIN 53714	%	--	3.1	--	2.5	--	2.4
Tensile Strength at Break	DIN 53455	MPa	58	49	54	44	47	42
Elongation at Break	DIN 53455	%	40	250	90	320	270	400
Flexural Modulus	DIN 53452	MPa	2485	1163	2279	1094	1712	789
Impact Strength-Izod Notched	ISO 180	KJ/m^2	10	35	19	60	30	90

a Ref. [16]
b Conditioned to equilibrium at 23°C/50% RH

involves two reactive component streams and the use of a dispensing unit for preparing the materials and filling the mold. Fig. 13 shows a processing schematic in which the separate streams, comprised of prepolymer dissolved in caprolactam (melting point 69°C) and catalyst dissolved in caprolactam, are maintained at a temperature of 100°C, mixed at low pressure, and conveyed to a mold heated to 160°C. In the mold the reaction proceeds typically to 97% completion in about 8 minutes. The nylon block copolymer crystallizes and the solidified product can then be removed from the mold with or without a cooling cycle.

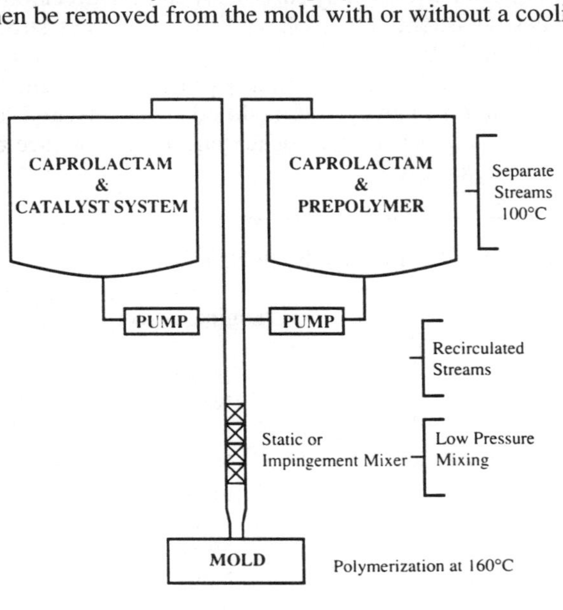

FIGURE 13 Processing schematic - RIM molding of nylon block copolymer

4.6.2 Tooling and Equipment Considerations

Existing rotomolding equipment can be used but may need modifications to permit the following [17]:

1. Adjustable speed in both major and minor axis up to 25 RPM
2. Axis rotation in both directions
3. Exact positioning of tool for filling, during the cycle and demolding
4. Heat exchanger (to preclude any flame source contact with caprolactam vapors inside the oven, to prevent explosion)
5. Improved temperature control. Cast aluminum and sheet steel molds can be used, but it is important that the mold be 'tight' in order to avoid material leakage at the parting line of the mold due to the low viscosity of the mixture of caprolactam, prepolymer, and catalyst at the 160°C process temperature. The maximum gap allowed in the parting line of the mold to prevent leakage is 0.05mm (0.002in.). Gaskets such as filled PTFE or silicone rubber may be used in the parting line to prevent leakage.

5. PROPERTIES OF MOLDED NYLONS

When properly processed and conditioned, nylons are tough, strong, abrasion resistant, self lubricating, have good fatigue resistance, and a high continuous use temperature of 120°C and 100°C for nylon-6 and nylon-12, respectively. The nylons are also noted for their chemical resistance. They are resistant to a wide spectrum of chemicals and are not stress-cracked or crazed by solvents such as hydrocarbons (gasoline, oils, greases), ketones, or esters. Alkalis have little effect, but strong acids and oxidizing agents will attack nylons. Table 7 summarizes the chemical effect of some classes of materials on nylons. This summary provides the chemical resistance of nylons at normal conditions; however, if modified chemicals are used or high temperatures are employed, it is best to specifically check performance by testing at the actual exposure conditions.

TABLE 7. CHEMICAL RESISTANCE OF NYLONS

Chemical Classification	Chemical Effect
Aromatic Solvents	None
Aliphatic Solvents	None
Gasoline, Gasohol,Oils	None
Weak Acids	None
Strong Acids	Attacked
Weak Alkalies	None
Strong Alkalies	None
Alcohols	None
Oxidizing Agents	Attacked

TABLE 8. EFFECT OF PLASTICIZER ON NYLON-11[a] AND NYLON-12[b]

Physical Property	ASTM Method	Unplasticized[c]	Plasticized[c]	
			Flexible	More Flexible
Nylon-11				
Tensile Strength at Yield, MPa[d]	D630	57	55	49
Elongation at Break %	D630	330	330	330
Flexural Modulus, MPa[d]	D790	980	490	345
Impact Strength				
Notched Izod, J/m[e] of Notch	D256	40	107	235
Nylon-12				
Tensile Strength at Yield, MPa[d]	D630	50	27	25
Elongation at Break, %	D630	200	250	300
Flexural Modulus, MPa[d]	D790	1400	600	400
Impact Strength				
Notched Izod, J/m[e] of Notch	D256	50	>500	>500

a Ref. 6
b Ref. 7
c Measured on dry-as-molded injection molded specimens
d To convert MPa to psi, multiply by 145
e To convert J/M to ft.-lbs./in,, multiply by 0.019

TABLE 9. EFFECT OF CAPROLACTAM PLASTICIZER ON ROTOMOLDING GRADES OF NYLON-6

Property	ASTM	Unplasticized[a]			Plasticized[a]		
	Method	-20°C	23°C	49°C	-20°C	23°C	49°C
Specific Gravity	D792	--	1.13	--	--	1.13	--
Tensile Strength at Yield,MPa[b]	D638	--	70	--	--	58	41
Elongation at Break, %	D638	--	20	--	--	160	--
Flexural Strength, MPa[b]	D790	134	97	38	--	68	28
Flexural Modulus, MPa[b]	D790	3276	2400	1400	--	1730	503
Falling Weight Impact Str., J[c]	D3029	27	165	213	21	173	275
Melting Point, °C	D789	--	215	--	--	215	--

a Measured on dry as molded samples cut from rotomolded parts.
b To convert MPa to psi, multiply by 145
c To convert J to ft.-lbs., divide by 1.375

5.1 Effect of Plasticizer

Plasticizers used in rotomolding grade nylon compositions are high-molecular-weight compounds of low volatility which have the ability to withstand high processing temperatures. In the case of nylon-11 and nylon-12, monomeric sulfonamides, such as N-(n-butyl) benzene sulfonamide, are used to increase flexibility and impact strength. The effect of plasticizers on molded properties of nylon-11 and

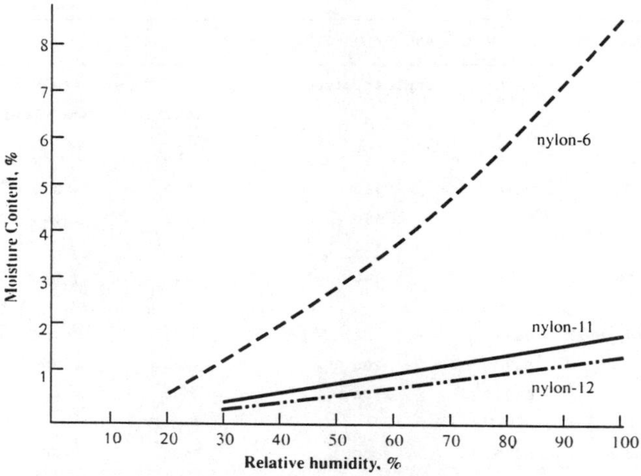

FIGURE 14 Moisture content vs. relative humidity. – – nylon-6, —— nylon-11, – •• nylon-12

12 is illustrated in Table 8. In the case of nylon-6, caprolactam which is not extracted or only partially removed from the polymerized product functions as an effective plasticizer. Unwashed nylon-6 polymer generally contains 8-10% by weight of unreacted caprolactam. Table 9 compares the rotomolded properties of unplasticized and plasticized nylon-6 resin. The flexural modulus of the plasticized nylon-6 resin is about one-half that of unplasticized resin at room temperature. An examination of the falling weight impact results reveals that, although the caprolactam plas-

TABLE 10. EFFECT OF WATER CONTENT ON PHYSICAL PROPERTIES [a]

Property	Nylon-6		Nylon-11		Nylon-12	
	Dry	50%RH	Dry	50% RH	Dry	50%RH
Tensile Strength at Yield, MPa	83	37	57	46	42	37
Elongation at Break,%	70	390	330	>330	>250	>250
Flexural Modulus, MPa	2700	1100	980	640	1420	980
Notched Izod Impact, J/m	65	215	40	–	69	78

a. Data compiled from trade literature

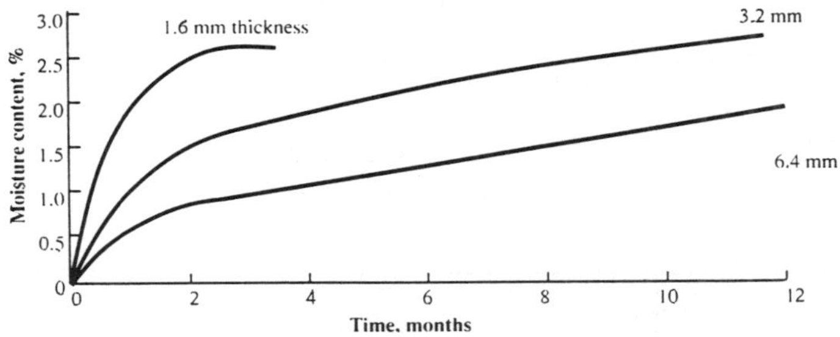

FIGURE 15 Moisture content vs. time for nylon-6, at 23°C, 50% RH

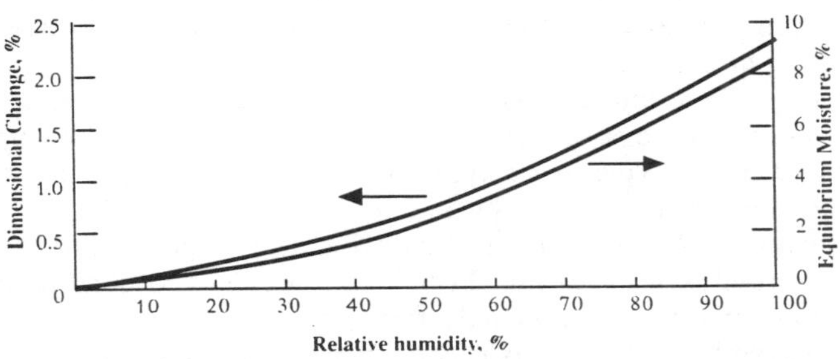

FIGURE 16 Moisture absorption and dimensional change - nylon-6 homopolymer resin

ticizer shows some benefit at elevated temperature, it has about the same impact as the unplasticized product at -20°C.

5.2 Effect of Moisture

With the absorption of moisture, molded nylon resins do not undergo degradation but become plasticized and undergo significant changes in mechanical properties. The moisturized molded product will develop improved toughness, but the designer should be aware that the resultant lower tensile strength and flexural modulus must be taken into account when designing a nylon part (Table 10). The amount and rate of moisture absorbed depends on the humidity (Fig.14), time and temperature of exposure, and the part thickness. Nylon parts exposed to ambient atmospheric conditions require extended time periods to achieve equilibrium moisture content. Fig. 15 illustrates the moisture absorption for nylon-6 parts at three different thicknesses exposed at 23°C and 50% RH.

Also, accompanying this moisture absorption is a predictable amount of expansion. For example, in Fig.16 is plotted the moisture absorption and dimensional change for a nylon-6 homopolymer resin. At 50% RH, a 3.2mm (0.125in.) thick

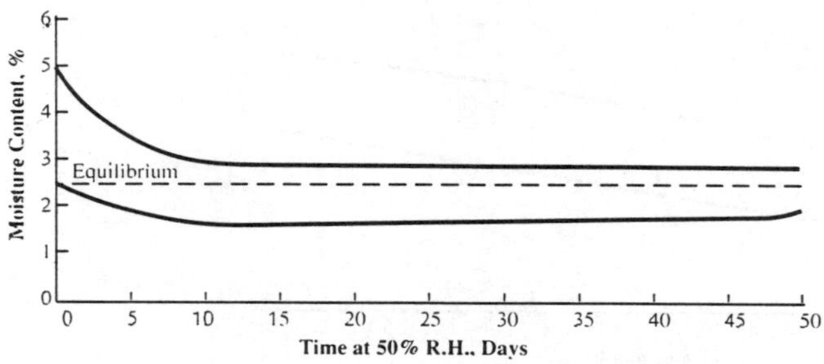

FIGURE 17 Moisture content vs. time after rapid conditioning - nylon-6

nylon-6 part in time will achieve an equilibrium moisture content of 2.7% and will expand about 0.18mm/mm (0.007in./in.) [18]. Nylon-11 at 50% RH will reach an equilibrium moisture content of 0.8% and will only expand 0.03mm/mm (0.0012in./in.). Although these dimensional changes are reversible (that is, a part that expands due to absorbed moisture will return to its original dimensions upon thorough drying) in general service, once nylon parts reach an equilibrium with their environment they remain relatively stable in dimension.

Some nylon applications having critical dimensional or impact values may require moisture conditioning prior to use. In this case accelerated conditioning methods are incorporated into the nylon manufacturing process. Accelerated conditioning is usually achieved by soaking parts in room temperature water for a period of time that will achieve the objective desired. If equilibrium moisture at 50% RH is the target, this is most expeditiously achieved by immersion of the part in water to reach a moisture content about twice that of the equilibrated value at 50% RH. This is required because the water will be absorbed primarily on or near the surface of the part, and when removed from water will lose moisture from the surface. This process is illustrated for a nylon-6 part 3.2mm (0.125in.) in thickness in Fig. 17. Two identical 12.7mm x 127mm bars were conditioned in water in two different moisture levels and then exposed to 50% relative humidity at room temperature. The bar conditioned to 2.6% moisture drops too far below the 50% equilibrium moisture value, while the bar conditioned to 5.0% moisture drops down in moisture but stays above the 50% RH moisture level of approximately 2.6%. The bar which dropped below the 50% equilibrium moisture value will require several months at 50% R.H. in order to reach the fully equilibrated value of 2.6% moisture.

5.3 Thermal Considerations

Nylons are considered good heat resistant materials because they are crystalline and therefore can withstand periods of exposure at temperatures almost to their melting points without a total loss of their basic mechanical properties of tension, flexure, compression, and shear strength. Fig.18 compares the shear modulus vs. temperature of a nylon-6 homopolymer with the amorphous polymer, polycarbon-

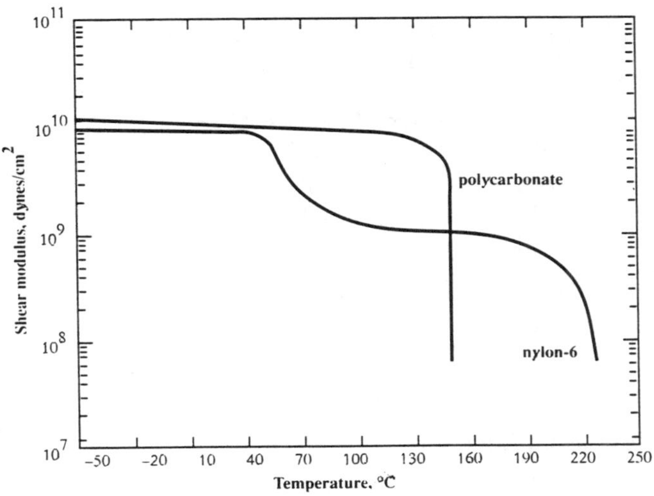

FIGURE 18 Shear modulus vs. temperature of nylon-6 and polycarbonate resins

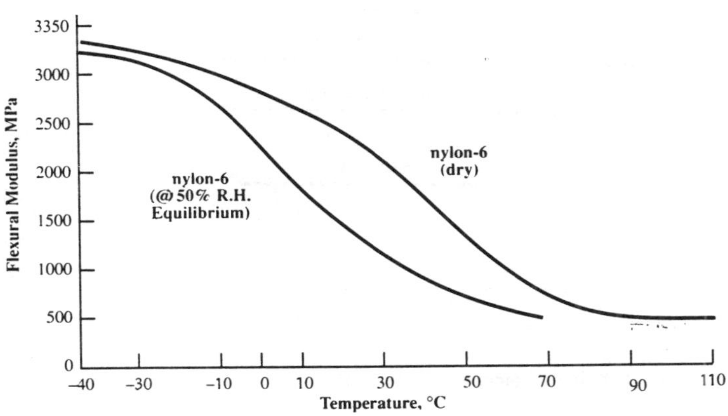

FIGURE 19 Effect of temperature and moisture on flexural modulus of nylon-6

ate, which loses all its mechanical properties above the Tg (glass transition temperature). This same ability of nylon to maintain some mechanical strength at elevated temperatures is illustrated in Fig.19 which plots flexural modulus vs. temperature for nylon-6 homopolymer tested dry and at 50% RH equilibrium. Flexural modulus is a measure of stiffness or rigidity of a fabricated part and is a most important short-term parameter when designing with engineering plastics.. Table 11 records the values of flexural modulus at -40° C, 23° C, and 80° C for the nylon resins.

TABLE 11.FLEXURAL MODULUS VS. TEMPERATURE FOR NYLON- 6, NYLON-11, NYLON- 12 [a]

Nylon Type	Flexural Modulus MPa [e]		
	-40°C	23°C	80°C
Nylon-6[b]	3310	2620	520
Nylon-11[c]	1324	980	186
Nylon-12[d]	1760	1300	260

a. Unplasticized grades of nylon
b. Ref. 19
c. Ref. 6
d. Ref. 7
e. To convert MPa to psi, multiply by 145

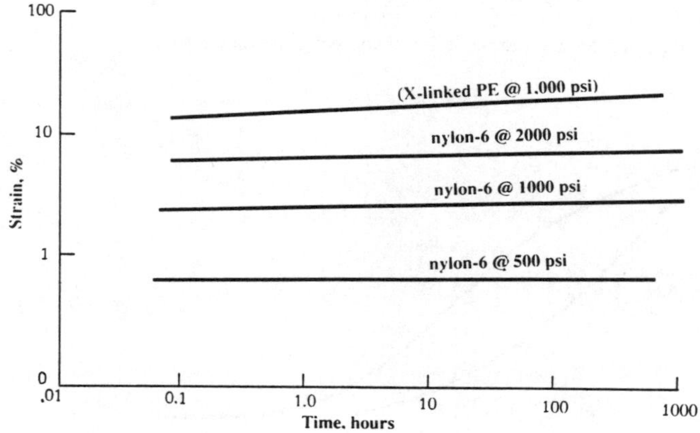

FIGURE 20 Tensile creep of nylon-6 homopolymer at 93°C

Another important property of nylons is creep resistance, which examines the long-term ability of a plastic to withstand a load without deformation or catastrophic rupture. Good creep resistance is the true ear-mark of an engineering plastic and is often essential in predicting part performance. All plastics tend to creep or to deform under constant, long-term load at a slow but continuous rate beyond the deformation predicted by their short-term moduli. Creep resistance is particularly important at elevated temperatures and is a key design property because of its time dependency. The tensile creep of nylon-6 homopolymer at 93°C (200°F) and at stresses which may be encountered in large tanks is given in Fig.20. At 1000 psi stress level, nylon-6 exhibits creep resistance substantially superior to a cross-

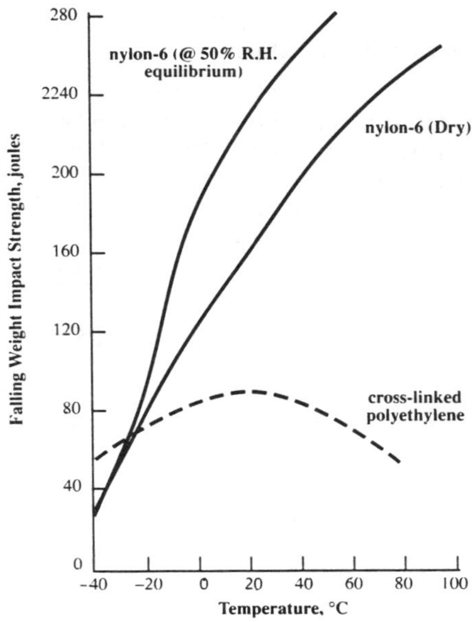

FIGURE 21 Effect of temperature and moisture content on falling weight impact strength of nylon-6 (102mm squares, 3.2mm thick)

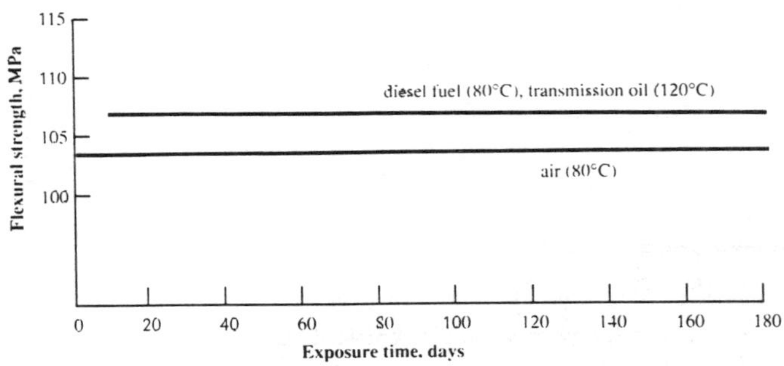

FIGURE 22 Effect of environmental aging on flexural strength of rotomolded nylon-6

linked polyethylene product.

While an increase in temperature or moisture content will reduce stiffness or rigidity, the toughness or impact properties of nylons are markedly increased, as can be seen in Fig.21. Impact resistance is one of the most important properties of nylon resins for rotomolding applications. The falling weight impact test as described in A.S.T.M. D-3029 is the preferred technique for measuring the toughness of a rotomolded part because this procedure closely simulates a part failure mode. Fig.21 also compares nylon-6 resin to cross-linked polyethylene. The data shows that the cross-linked polyethylene material has better impact than nylon-6 at -40°F, but substantially lower impact at elevated temperatures.

TABLE 12. CONTINUOUS USE TEMPERATURE RATINGS FOR MECHANICAL PROPERTIES WITH IMPACT-UNDERWRITERS LABORATORIES[a]

Resin Type	Thickness, mm 1.47	3.05	6.10
Nylon-6	105°C	105°C	105°C
Nylon-11	65°C	65°C	65°C
Nylon-12	70°C	70°C	—

a Ref. 20

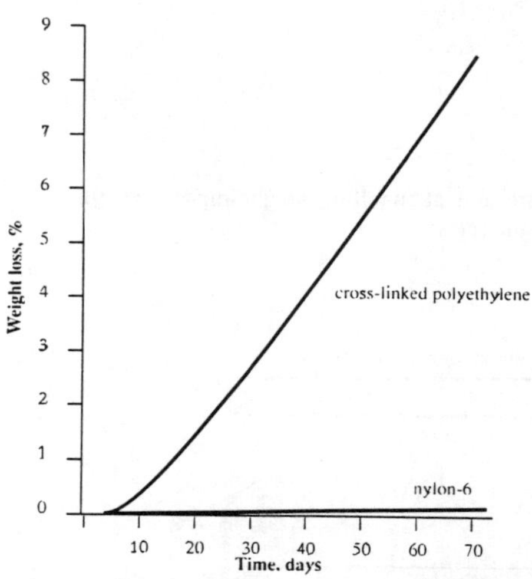

FIGURE 23 Room temperature gasoline permeation of rotomolded tanks - nylon-6 vs cross-linked polyethylene

Nylons will degrade when exposed to high temperatures and a loss in property values will occur. The degree of degradation will depend on the time and temperature of exposure. The Underwriters Laboratories' continuous use temperature ratings for mechanical properties with impact are listed in Table 12 for heat stabilized nylon resins at thicknesses of 1.47,3.05,and 6.10mm [20].

For heat stabilized nylon resins, suppliers usually specify continuous use temperatures of 120°C, 100°C, and 100°C for nylon-6, nylon-11, and nylon-12, respectively. An example of the heat resistance capability of nylon-6 under different environments can be seen in Fig.22. Here specimens cut from rotomolded parts were exposed for 180 days at 80°C in diesel fuel, at 120°C in transmission oil, and as a control at 80°C in an air-circulating oven. In Fig.22 flexural strength vs. exposure time is plotted for each of the environments. The flat curves obtained show that there was no decrease in flexural strength after about 6 months' exposure to the different environments.

5.4 Permeability

Nylons offer low permeability to oxygen which makes them suitable for any air-tight packaging. They also have low permeability to gasoline, diesel fuels, xylene, and many other solvents at ambient or elevated temperatures. Fig.23 illustrates the gasoline permeation obtained in identical tanks rotomolded in a cross-linked polyethylene (CL-PE) resin and in nylon-6 resin. The results show that after 70 days' exposure at room temperature (23°C), there was over 8.5% weight loss through the CL-PE tank and less than 0.1% through the nylon-6 tank. At elevated temperatures the superior permeation resistance to gasoline of nylons compared to polyethylene would be expected to be even more pronounced.

6. FINISHING AND MACHINING OF NYLONS

Like polyethylene resins, after rotomolding processing, nylon resins will require some finishing and possibly machining before the molded product can be shipped to the end-user. The usual finishing operation is the removal of any mold parting-line flash which occurred during processing. Since nylons have a greater hardness and are substantially tougher than polyethylene, it is usually more difficult to remove flash from the molded part. Flash removal from the nylon part is less difficult if it is removed while the part is still hot right out of the mold. Removal of flash is accomplished using a variety of sharp-bladed tools (scrappers) held so that they pass over the flash without digging under the surface of the part.

In many rotomolding applications it is necessary to saw, rout, drill, or tap the molded part to prepare it for its end-use function. In all types of machining operation on nylon parts, it is important to minimize heat generation. The heat conductivity of nylon is much less than metals, and as a result heat generated by friction between the nylon and the metal cutting tool will be mostly absorbed by the cutting tool [21]. The minimal heat conducted into the nylon plastic will not transfer through the part cross-section, and as a consequence the temperature of the surface layer will rise significantly. If this surface heat is not controlled it can cause gumming and discoloration, and produce an unsatisfactory part. The proper selection of sharp cutting tools, cutting speeds, and water cooling or an air blast that removes chips from the cutting area will help prevent overheating.

6.1 Sawing

Power band sawing of nylon is widely used for making irregular or curved contours as well as straight cuts. For a material thickness of less than 0.5 in. (12.7mm), the best results are obtained using high carbon steel blades with a precision-type tooth form having 10-14 teeth/in. and a band speed of 3000 fpm. Since the dissipation of heat is also a problem in band sawing, it is important to use the proper feed rate. Because most sawing operations are hand-fed, the feed rate is best determined through experience by trial and error [22]. Too high a feed rate will result in heating of the blade which will cause gumming of nylon and loading of the saw teeth. In addition to the above factors, to ensure the best cutting performance on nylon, the band saw should be maintained with a sharp saw blade.

TABLE 13. DRILLING RECOMMENDATIONS FOR NYLON [a]

Drill Speed m/min.	Feed, mm/revolution Nominal Hole Diameter							Tool Material Grade
	3mm	6mm	12mm	18mm	25mm	25mm	50mm	
30 to 76	.05	.10	.13	.15	.20	.25	.30	52,53 (ISO) M10,M7,M1 (AISI)

a Ref. 24

6.2 Drilling

With due consideration given to their thermal characteristics and high toughness, nylon parts can be easily drilled. By minimizing heat generation and ensuring chip removal, nylons can be drilled satisfactorily with conventional twist drills. However, more rapid and better results are obtained with special drills, designed for plastics, having highly polished, deep flutes to accommodate removal of chips and reduce frictional heat. The drill geometry recommended for nylon resins calls for a helix angle of 17°, point angle of 70°-90° and a clearance of 9°-15°. The other drilling parameters for nylon at various nominal hole diameters can be found in Table 13 [23]. To maintain quality and ease of drilling, frequent sharpening of drill points is essential.

6.3 Tapping

The same tool material grades recommended for drilling nylon are suitable for tapping operations. In order to develop less frictional heat, finish ground and polish flute taps are recommended. Because of the elastic recovery of plastic materials, oversize taps should be used with nylon. For tapping nylon the pitch diameter should be 0.13mm (0.005 in.) oversize, unless a tight fit is desired. For nylon, taps having 3 or 4 flutes should be used at cutting speeds of 23 to 38 m/min. (75 to 125 ft./min.) [24].

6.4 Routing

For routing work on nylon parts, a 7 Hp router capable of 18,000-20,000 rpm is generally used with a double-fluted high speed steel or carbide-tipped bit. As with all machining procedures with nylons, the router bits must be kept sharp and a guide should be used to ensure good cutting performance. Computer numeric control (CNC) systems are available which can be set up to perform many routing operations automatically.

7. NYLON APPLICATIONS

The many valuable physical and thermal properties of the nylon resins have found applications in a variety of markets. Rotomolded nylon resins have replaced metal parts because of one or more of the following cost/performance factors:

(1) ease of manufacturing
(2) ability to manufacture seamless complex shapes
(3) corrosion and chemical resistance
(4) explosion resistance (no sparks produced when impacted by some puncturing metal object)
(5) better vibration and fatigue resistance
(6) better abrasion resistance
(7) weight and cost reduction

Nylons have replaced some metal parts in air-intake ducting for trucks by virtue of their high heat resistance, high strength, weight reduction, and resistance to fuel and oil. Nylon abrasion resistance has resulted in some agricultural equipment parts such as hoppers and air-seeder components being rotational molded. In many of these applications, the rotomolding process provides the most economical means of producing relatively low volume parts. Unlike the typically small size parts that injection molding processors are accustomed to producing, nylon rotomolded products such as 5300 liters (1400 gal.) chemical storage tanks have been molded weighing as much as l66kg.(365 lbs.) In fact, there is no material limitation on size of rotomolded nylon parts.

In one fairly large application, nylon-6 is being used in a 587 liters (155 gal.) fuel tank for the Bradley M-2 Infantry Fighting Vehicle. The fuel tank was designed to utilize every available amount of space within the vehicle due to severe restrictions of space [25]. As a result, this fuel tank design is most complex and could not be molded by any process other than rotational molding. The tank measures approximately l.5m x l.7m x 0.4m (5 ft. x 5.5 ft. x 1.3 ft.) with a nominal wall thickness of 0.312 in. (7.9mm) and with a molded-in aluminum spool at its center. Nylon-6 was chosen for this application because the fuel tank had to withstand fuel temperatures as high as 250°F. Besides being impervious to diesel fuel, the nylon provided the necessary flexural modulus so that it maintained its shape with the weight and heat of the fuel. In addition, the fuel tank had to withstand vertical-shock loads of ten times the force of gravity which are encountered when the vehicle becomes momentarily airborne in traversing rough terrain at 64km/h (40 mph) speeds. Aluminum, which was initially considered for use in this tank, was ruled out because of the impracticality of this complex shape being fabricated with welded metal. Anticipated were severe problems of cracks in welded joints caused by shocks and vibrations during vehicle operation. Also, the rubber bladder required in the aluminum version of the fuel tank would have cost more than the entire tank made out of nylon. Cross-linked polyethylene was considered, but it would have no rigidity at temperatures above 200°F and it lacked the resistance to permeation by diesel fuel.

Nylon-6 has found use in storage containers for solvents. Tanks 380 to 1890 liters (100 to 500 gallons) are being used in the dry cleaning industry for storage of perchloroethylene. The excellent resistance of nylon-6 to perchloroethylene made it a logical choice for this application, which developed as a result of corrosion problems encountered by users of epoxy-lined steel tanks. Since polyethylene was unacceptable due to severe swelling, distortion, and lack of adequate permeation resistance to this chemical, and since stainless steel tanks were too costly, nylon-6 was the cost/performance choice for this application.

Significant cost reductions have been achieved by equipment manufacturers with the converstion of hydraulic oil reservoirs from fabricated steel construction to rotomolded nylon tanks. Besides savings on fabrication, in many cases novel configurations were possible which maximized tank capacity. In one lift truck application, problems of corrosion and rust associated with steel tanks were also eliminated.

Nylon rotomolding resins, as engineering thermoplastics, have substantial advantages over polyethylene which make them the best choice for applications with one or more of the following requirements:

(1) better heat resistance
(2) greater tensile strength and flexural modulus
(3) creep resistance
(4) resistance to permeability by gasoline and other fuels
(5) chemical resistance to aromatic solvents
(6) superior environmental stress crack resistance
(7) abrasion and wear resistance
(8) ease of paintability (nylon can be painted like metal without any surface treatment; polyethylene must be flame-treated prior to painting)
(9) toughness, especially in molded-in or machined threads

PA range of applications which often require some of the above special qualities of the rotomolding grades of nylon are listed below:

(1) tanks – for gasoline, diesel-fuels, solvents etc.
(2) ducting – for automobile pollution control, air-intake, exhausts, ventilating systems
(3) hydraulic oil reservoir tanks
(4) oil-displacement tanks
(5) tanks for use in vacuum systems
(6) automobile dash board trim moldings
(7) housings for medical and marine equipment
(8) abrasion resistant conduits
(9) mandrels for filament winding
(10) inner liner for filament wound high pressure tanks.

REFERENCES

[1]. Floyd, D.F., 'Polyamide Resins', Reinhold Publishing Corp. 1966, p.3 (book)
[2]. US. Pat. 4,548,779 (Oct.22,1985) A.H. Steinberg and F. Petruccelli (to Allied Corp.) (patent)
[3]. Puffr, R. and Kubanek, 'Lactam-Based Polyamides', Vol.II, CRC Press, Inc., 1991, p.174-175 (book)
[4]. CAPRON Nylon Homopolymers for Molding and Extrusion, CL-100-685, Allied Corp., Morristown, N.J. (trade literature)
[5]. Zytel Nylon Resins, General Guide to Products and Properties, E-42267, E.I. du Pont de Nemours & Co., Wilmington, De. (trade literature)
[6]. Design Guide to a Versatile Engineering Plastic, Rev. 4/84, Rilsan Corp., New Jersey (trade literature)
[7]. Vestamid Polyamide 12, Huls Chemische Werke Huls AG, Germany, Dec., 1979. (trade literature)
[8]. Vydyne Engineering Plastics Composite Data Sheet, No. 6313 D, Monsanto Plastics and Resin Co., St-Louis, Mo. (trade literature)
[9]. Technyl Polyamides, Properties Des Principales Formules Technyl, Rhone-Poulenc Specialités Chimiques, 1982 (trade literature)
[10]. U.S. Pat. 4,508,675 (Apr.2,1985) P.P. Salatiello and F. Petruccelli (to Allied Corp.) (patent)
[11]. Ref. 1, p.61 (book)
[12]. Kohan, M.I., 'Nylon Plastics', John Wiley & Sons, New York, 1973, pp 28,29 (book)
[13]. Ref. 10 (patent)
[14]. Progelhof, R.C., Empirical equation developed for Allied Corp. at New Jersey Institute of Technology, 1978 (report)
[15]. Sweeney, M.F. 'Reaction Injection Molding Machinery and Processes', Marcel Dekker, Inc., 1987, p.l03 (book)
[16]. Nyrim for Rotomoulding, Publ. No. 43-5 (E) RIM, DSM RIM Nylon Inc., The Netherlands and Georgia, USA, 1990 (trade literature)
[17]. Ref.16, pp. 4-6 (trade literature)
[18]. Dimensional Considerations with CAPRON Nylon, Engineered Plastics, Allied-Signal Inc., Morristown, N.J., 1990 (trade literature)
[19]. CAPRON Nylon 8280 HS for Rotational Molding EPC-104-85, Allied Corp., Morristown, N.J., 1985 (trade litrature)
[20]. Recognized Component Directory, Underwriters Laboratories, Inc., Vol.II, Northbrook, Il., 1990 (book)
[21]. SPI Plastics Engineering Handbook, Fifth Ed., Van Nostrand Reinhold, 1991, p.666 (book)
[22]. Ref. 21, pp. 684,685 (book)
[23]. Machining Data Handbook, 3rd. ed., Machinability Data Center, Metout Research Associates Inc., Ohio, 1980, p.49 (book)
[24]. Ref. 21, pp. 673,675 (book)
[25]. Anon., 'Nylon-6 Fuel Tank Pronounced Combat Ready', Plastic Design Forum, March/April, 1981 (journal article)

CHAPTER 4
Rotational Moulding Machines
W.H. Covington, Jr.

1. INTRODUCTION
Rotational molding machines have undergone significant concept changes from the 1950's to present. This chapter will describe the different types of machines and their general operation that are used to produce rotationally molded parts. Where possible, we will outline the advantages of using each type of machine design.

2. THE ROTATION PROCESS
The rotational molding process uses three parameters to produce parts: time, temperature, and mold rotation. The rotation process is accomplished by rotating the mold about two axes — a vertical and a horizontal axis.

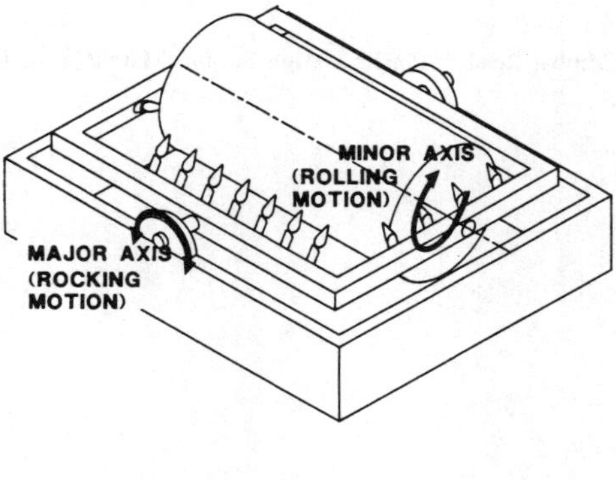

FIGURE 1 Open flame rock and roll machine

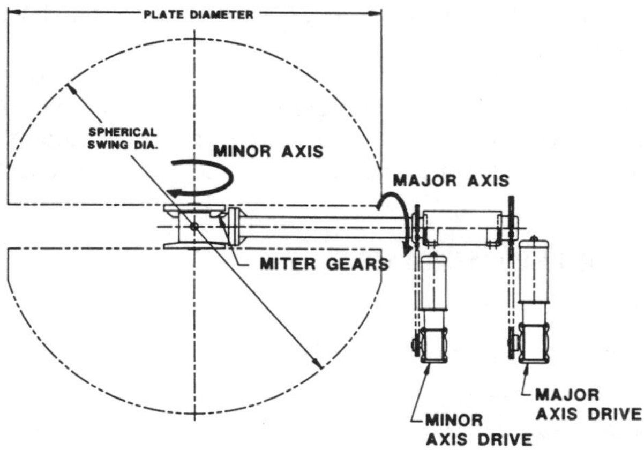

FIGURE 2 Straight arm

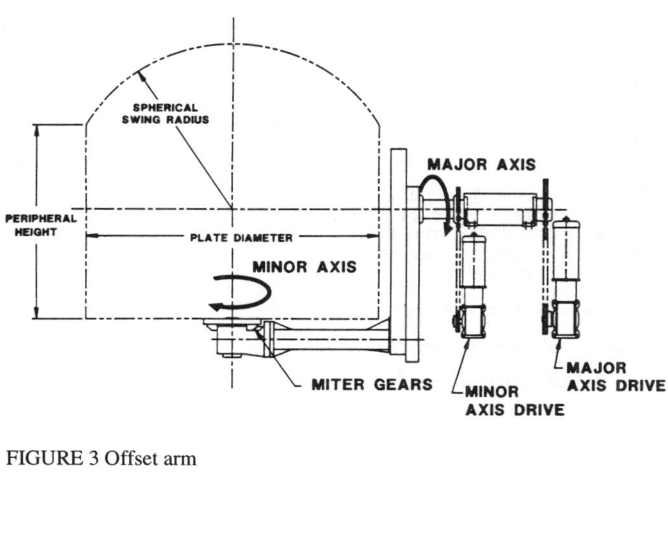

FIGURE 3 Offset arm

The early rotational molding process used a "rock and roll system", as illustrated in Figure 1. The mold was placed on a horizontal axis with a power driven roll providing the rolling axis similar to a barrel roll. Simultaneously, the entire framework on which the rolling action takes place was equipped with a mechanical or hydraulic drive system that enabled the frame to be tilted 45° to the left and 45° to the right; hence rock and roll.

The most conventional method of rotation used is "biaxial rotation" as illustrated in Figures 2 and 3. Figure 2 is a straight arm, a rotation member which protrudes horizontally from the machine center, with molds positioned on either side of the major axis. The major axis is rotated by direct connection to the gear motor, while a

ROTATION RATIO FOR TYPICAL SHAPES

RATIO	SHAPES
8 to 1	Oblongs (Horizontal Mounted) Straight Tube (Horizontal Mounted)
5 to 1	Some Defroster Ducts
4.5 to 1	Balls or Gloves
3.3 to 1	Any shape showing overlapping lines of rotation at 4 to 1
4 to 1	Cubes - Balls - Odd Shapes Rectangular boxes - Horses with bent legs
2 to 1	Rings, Tires, Balls Any rectangle which shows two or more thin sides when run at 4 to 1 Picture Frames - Mannequins Round Flat Shapes Horses with straight legs Auto Crash Pads (Vertical Mounted)
1 to 2	Parts which should run at 2 to 1 but show thin side walls
1 to 3	Flat rectangles (gas tanks - suit cases - tote bin covers)
1 to 4	Tires - Curved Air Ducts Pipes Angles - Flat rectangles Balls whose sides are thin at 4 to 1 ratio (Vertical Mounted Cylinders)
1 to 5	Cylinders (Vertical Mounted)

Horizontal Mounted - Parallel to major axis.

Vertical Mounted - Perpendicular to major axis.

FIGURE 4

pair of right angle miter gears, turned by an internal shaft, provides the minor axis rotation. The minor axis power source is also delivered by a power source. By varying the speed of each of the independently driven axes, the operator can obtain the "rotation ratio". Figure 4 shows a table which lists the ratios between the major axis speed and minor axis speed that may be used for different sizes and shapes of product. This ratio allows proper distribution of the material on the internal surfaces of the mold as it is heated. While the chart for Figure 4 shows typical speeds and ratios that might be used for producing parts, it is often left up to the operator to determine the exact speed and ratio for each particular mold. The shape of the mold and the polymer material to be used often dictate certain speeds and ratios.

Figure 3 illustrates an offset arm; this device enables the molder to put a large mold in the same oven volume that is available to the straight arm. Typically, one can put fewer molds on an offset arm than on a straight arm; however, the offset arm allows larger pieces to be produced than does the straight arm. Rotation is achieved in the same way with power being applied to the major (horizontal) axis for rotation and, through an inner shaft and a set of miter gears, rotation is achieved in the minor (vertical) axis. The same guidelines for speed and rotation apply for offset arms as for straight arms.

When we refer to the "mold swing," we are speaking of the volume developed by the swing of the major (horizontal) axis and minor (vertical) axis in the oven and cooling chamber. For different sizes of machine mold swing varies. As the

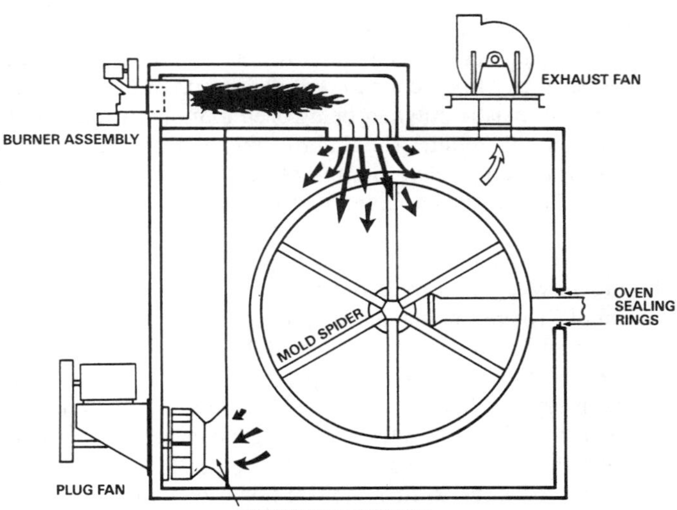

FIGURE 5 Recirculating hot-air oven diagram

space inside the swept volume of the oven is increased, the weight capacity of the arm usually is increased because larger molds weigh more and carry more powder weight. For example, a machine designated a 2000/95 machine, would mean that the arm weight is rated at 2000 pounds capacity and 95" of spherical swing. Figures 2 and 3 show illustrations of the spherical swing for the straight or offset arms, respectively.

Heating Methods

Rotational molds may be heated by either an *open-flame method,* a *hot air recirculating oven method,* or by a *hot-oil jacketed* mold system. The most used system is a recirculating hot-air oven. In this system, illustrated in Figure 5, a positive displacement circulating fan distributes air through a system of ducts into the swept volume of the oven. The capacity of the fan (cubic meters of air per minute), will determine the number of air changes per minute. On contemporary machines, air should be changed in the oven approximately 25-30 times per minute in order to provide an effective heating for the molds. Direction of the air in the oven is generally caused by the directional louvers so that no "dead spots" are created. The static pressure capability of the fan system provides force to push the air over the mold and provide the scrubbing action of the hot air on the mold. The absorption of the heat by the mold transmits through to the powder to create the molded part.

The medium for heating hot-air ovens may either be natural gas or oil with a modulating burner. In some cases, electric heaters are used to generate the hot-air environment.

The regulation of air temperature in the swept volume of the oven is controlled by sophisticated electronic temperature control devices.

The time that the mold remains in the oven is known as "oven residence time". The oven residence time necessary to cure a part will depend upon the thickness of the wall of the part, the type of plastic polymer being used to mold the part, and the conductivity of the metal of the mold. Aluminum, with a higher conductivity, allows heat to transfer from the air stream to the mold and the powder at a much faster rate than does steel. Thinner gauge aluminum helps to increase the rate of heat transfer.

On the earliest rock and roll machines there was no heated oven; an open-flame method was used (Figure 1) whereby a manifold of gas jets was placed to evenly heat the mold. As the mold rotated about the major (rolling) axis, the heat was imparted directly onto the mold surface, and transferred through to the plastic. This machine was inexpensive to manufacture, but the operating costs were significantly more than the closed oven type of heating system. All of the thermal energy not imparted to the mold went into the atmosphere, creating increased temperature in the work environment and the loss of energy. Open flame machines are still used for very large tanks that are too large to fit in conventional recirculating ovens or where the quantity of tanks is so small as to not justify the expense of building a large oven. In some instances, modern safety laws have out-moded the use of open flame systems.

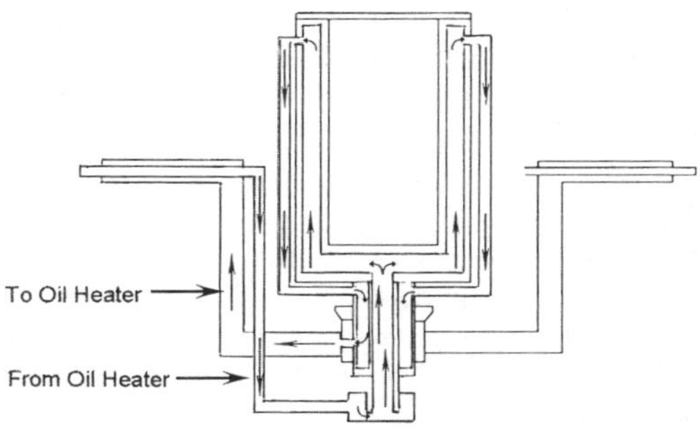

FIGURE 6 Hot oil jacketed mold system

FIGURE 7 Rocking oven machine

The hot oil jacketed mold system (Figure 6) was one of the earliest systems used for rotational molding, and actually provides one of the most efficient means of transmitting heat to the plastic part. Whereas in the closed oven with recirculating hot air, you have heat losses due to the extra volume in the oven not filled by the mold and losses through the skin of the oven as well as through discharge each time the oven doors are opened and closed, the jacketed mold maintains the mold temperature very close to the temperature of the hot oil being used. Therefore, the hot oil system generally uses a lower temperature for molding since the oil is in direct contact with the mold and imparts the heat energy very quickly. The difficulty of using jacketed hot-oil molds is that the expense of the molds is considerably more than those used in the other types of heating. Early hobby horses and industrial tanks were made using a hot-oil method.

3. ROCK AND ROLL MACHINES

We have earlier discussed open-flame rock and roll machines for rotationally molding products. As illustrated in Figure 1, an open-flame rock and roll machine is an economical way to begin the rotational molding process, particularly for large, cumbersome parts. In Figure 1 a large cylindrical tank is rotated (rolling) about the major axis and tipped left to right about the minor axis (rocking) in order to allow the powder to distribute in the mold. During this process, the major axis (rolling) is always rotating at a much faster speed than the tilting axis.

An alternative to the open-flame rock and roll machine is the "rocking oven machine" (Figure 7). The mechanical principles here are still the same. The mold is rolling about the main axis and is rocking 45° to the left and right about the minor axis. The rocking oven machine is significantly more sophisticated in that it uses recirculating hot-air systems to reduce energy losses and to create the proper high scrubbing heat transfer from the air to the mold. All the advantages of the recirculating hot air oven are enjoyed with this machine while still giving the capability of producing long, slender objects such as industrial tanks, kayaks, canoes, lampposts, and other engineered products.

In the cooling station, of the rocking oven machine, the material is already set in a fused state in the oven. Therefore, it is not necessary to rotate the part with a biaxial system; here the mold is rotated only on the rolling (longitudinal) axis while cool high pressure air is blown across the mold to remove heat. In order to increase the rate of cooling, a water mist or water spray may be introduced later in the cooling cycle. While both the open-flame rock and roll system and the rocking oven machines have the advantage of being able to produce large parts economically, on the disadvantage side they are slow and cumbersome because the system is not very automated.

4. BOX OVEN MACHINES

This fundamental machine has been used for years in the production of play balls, baby doll parts, and automotive components such as shifter boots or door pulls. Because of the nature of the operation of the machine, it is generally relegated to

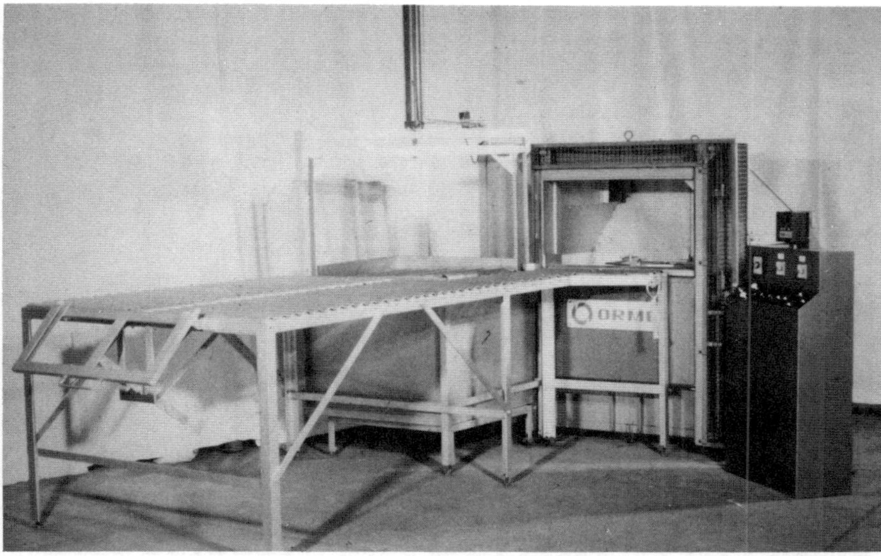

FIGURE 8 Box oven machine

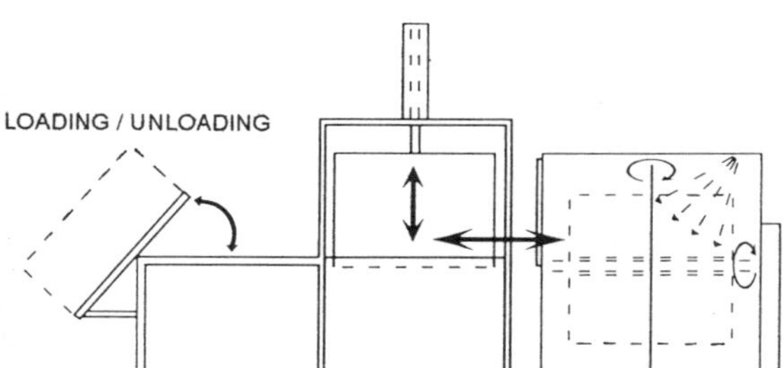

FIGURE 9 Box oven operation principles

FIGURE 10 Shuttle-style machine

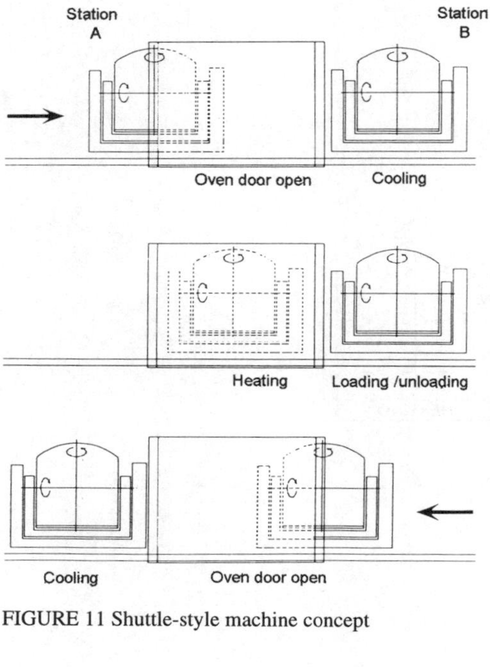

FIGURE 11 Shuttle-style machine concept

molding liquid plastisol PVC parts, but there are circumstances where it can be used for small, hollow articles of polyethylene or other thermo-plastic materials. Figures 8 and 9 illustrate the Box Oven Style machine.

The box oven machine uses biaxial rotation in the oven; heat in the oven is supplied by a gas or oil fired burner and a recirculating fan system. The cooling system is quite simplistic with the mold assembly placed on a roller mechanism that is plunged, by means of an air cylinder, into a water bath and held there until the optimum temperature of the part has been reached. The air cylinder retracts, raising the mold assembly back-up to platform height. The mold assembly is shifted to the loading/unloading station for opening the mold and removing the parts and recharging the mold.

Because of the manual labor involved in the transferring of the molds from the loading station to the oven, back to the cooling station, and then to the unloading station, this machine has been generally outmoded in all but some primitive molding operations. The machines have been replaced with modern carousel types.

5. SHUTTLE-STYLE MACHINES

Almost from the beginning of rotational molding, shuttle-style machines were utilized to produce large tanks and containers. Even today, one would be pressed to find a simpler, more cost-effective way of producing large parts than the venerable shuttle-style machine illustrated in Figures 10 and 11. These machines are simple to manufacture, and have reasonably low costs for manufacturing.

The shuttle machine operation concept is as follows: Shuttle A is transported into a hot-air oven, the oven door closes, and rotation on both axes is commenced. At the completion of the preset oven time, Shuttle A is removed from the oven and begins its cooling process, in a controlled water/air environment. After Shuttle A is clear of the oven, Shuttle B enters into oven to begin its curing cycle. While shuttle B is in the oven, the mold on shuttle A is cooled, the part is removed. The mold is reloaded for the next cycle.

The advantages of shuttle-style machine are:

1. The ability to mold large and small parts economically.
2. Simple to operate with reasonably low maintenance.
3. Simplistic design minimizes floor space consumption.

The general configuration of the shuttle machine allows easy access to the molds. The machine operation is not as time-paced as automated carousel machines, thereby affording the operator the opportunity to manage large parts and molds.

Shuttle machines may be initially provided with one mold shuttle and one oven. Later, the machine can be outfitted with a second shuttle to improve the productivity of the machine. This versatility allows the shuttle machine to be a low cost entry level molding process.

6. CLAMSHELL ROTATIONAL MOLDING MACHINES

This unique machine design was developed in the early 1980's to provide machines with low initial cost, low floor space requirements, low labor content, and good quality rotational moldings. In this style machine, the heating cycle is conducted first in a hot-air oven; after the cure is complete, the burner is extinguished; the same circulating air fan is used to cool the product. In the cooling cycle a bypass duct allows plant air to be ducted into the oven chamber to provide the cooling air. Later in the cooling cycle, water spray is introduced into the oven to cool the mold.

In this molding process biaxial rotation is provided by variable-speed gear-motors for the major and minor axes. The machines are available with either straight arms or offset arms, which enables the handling of a variety of small or large molds.

The advantage of the system is that the heating and cooling are all conducted in one closed environment. This minimizes the amount of floor space.

A disadvantage of the clamshell machine might be in its energy consumption since the oven is heated from ambient temperature to molding temperature and then cooled back down to ambient temperature.

7. VERTICAL STYLE MACHINES

This unique rotational molding machine is typically available in two types: 3-arm and 6-arm machines. The 3-arm machine is illustrated in Figures 12 and 13, and the 6-arm machine is illustrated in Figures 14 and 15.

Both the 3-arm and 6-arm machines are used to make doll parts, toys, play balls, automotive components, and, occasionally, technical products such as light globes, road furniture or other similar items.

The 6-arm machine is particularly useful for vinyl plastisol parts because the process typically requires a low temperature in the first oven zone to get the material to flow over the mold surface; the second and third oven stations are used to gel and complete the cure.

Like the clamshell machines, the vertical machines are self-contained machines utilizing a minimum amount of floor space. Unlike the clamshell machines, the vertical machines have a separate loading/unloading station, ovens, and cooling stations. These machines operate much like a "ferris wheel"; i.e., indexing is in a vertical plane. The 3-arm version operates with the operator in front; the machine indexes one mold vertically from the loading station to the oven station while another mold is simultaneously being cooled, and the third mold moves to the loading station. In the case of the 6-arm machine, the total heating cycle for the part is divided into three equal steps. Because there are three separate oven chambers, the part (mold) spends one-third of the required oven time in each of the three ovens. The machine is also provided with two cooling stations, so cooling time for a part is divided into two equal cycles. Because of very short dwell time in each section, the operator is presented with finished parts in short intervals.

The limitation of the vertical machines is not versatility or productivity, but the size of part that can be molded. Maximum mold swing is generally limited to 900mm to 1200mm.

On both versions of the vertical machine, each arm is provided with variable-speed major and minor axis drives. Heating is accomplished with a gas fired forced-air oven. The cooling medium is a normal air and water mix.

FIGURE 12 3-arm vertical machine

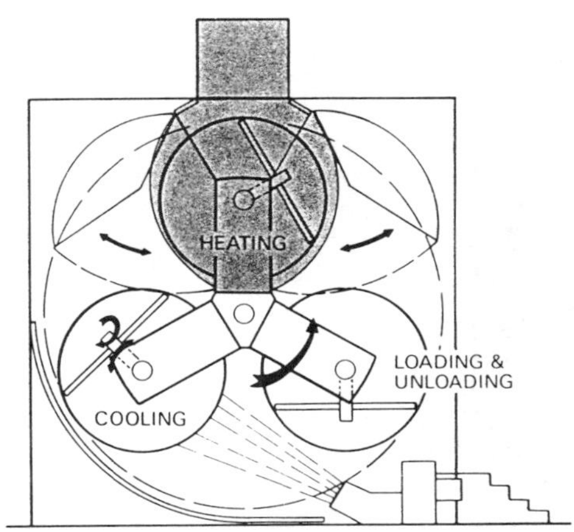

FIGURE 13 3-arm vertical machine: cycle diagram, *side elevation*

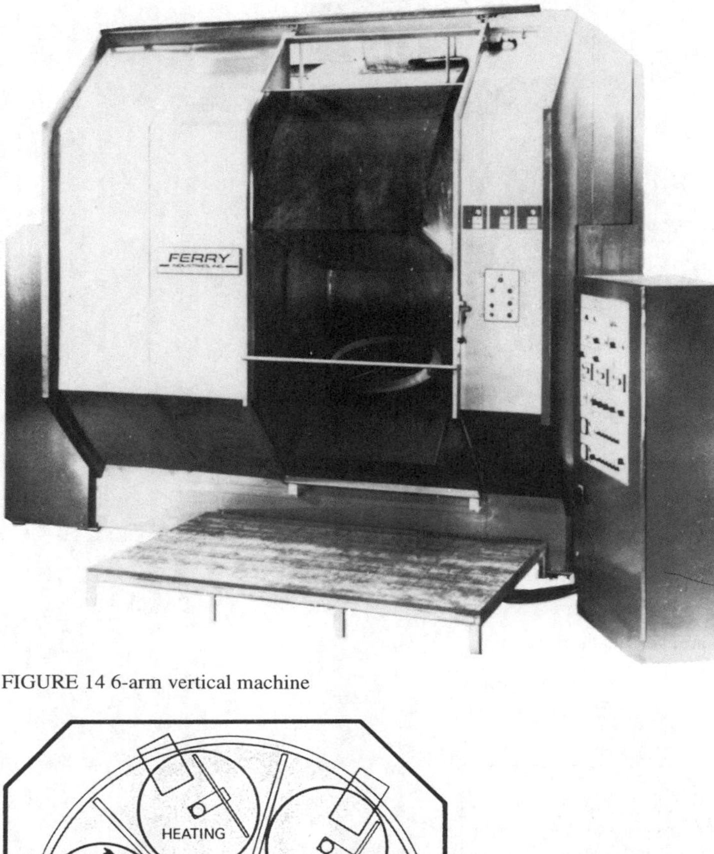

FIGURE 14 6-arm vertical machine

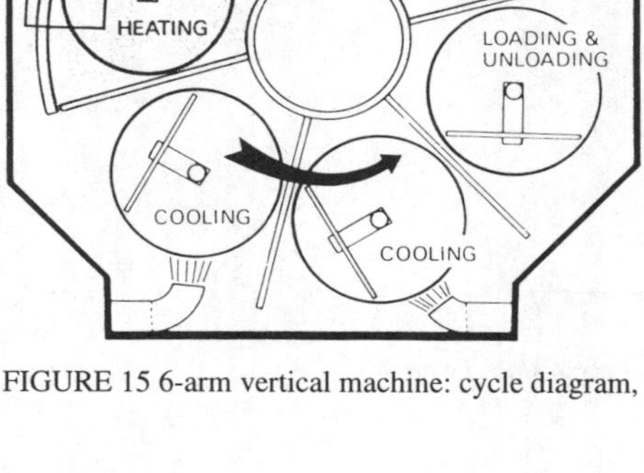

FIGURE 15 6-arm vertical machine: cycle diagram, *side elevation*

8. FIXED-ARM TURRET MACHINES

This popular concept of rotational molding machine was first introduced in the late 1950's or early 1960's. Machine sizes range from small machines with 40" swing diameter to machines as large as 150" swing diameter. The fixed turret style machine dominates the usage in the field today. These machines are cost-effective, highly productive, and easily maintained.

A fixed-arm turret machine (Figure 16) is used in applications where the oven and cooling cycles require relatively equal process time and where the load/unload cycle can be easily managed. The 4-arm version is required where one of the process cycles, *loading* or *heating* or *cooling*, requires more time than any of the other cycles.

The basic fixed oven machine has three stations — an oven station, a cooling station, and a loading/unloading station (Figure 17). The ovens are usually forced hot-air recirculating ovens with power operated doors on the entry side and on the exit side. The cooling chambers are equipped with doors on the entry and exit sides and are equipped to provide a cooling medium of forced air and water spray or a mixture of both. The loading station provides a position where the machine operator can open the molds to remove the parts and then refill the molds for the next

FIGURE 16 Fixed-arm turret machine

cycle. In this station, the arm can be semi-automatically rotated to access the mold from the opposite side of the arm.

All three of the arm assemblies are mounted on a fixed turret assembly, at 120° angles. When one arm must be indexed, they must all move simultaneously. Consequently, one work station dictates when the machine must index. Normally, the oven is the most critical part of the cycle, hence it becomes the governing factor in the indexing of the machine.

In recent years, it has been found that the cooling portion of the cycle may be as significant as the heating portion of the cycle. In order to afford more opportunities for cooling, the 4-arm machine was developed. The 4-arm fixed-turret machine provides four arms at equal 90° spacing on the fixed-turret. In this configuration, as shown in Figures 18, 19, 20, and 21, the most modern 4-arm machine can provide process cycles that allow for double loading/unloading stations requiring two operators for loading and unloading, or the machine may provide two cooling stations for situations where the cooling cycle needs to be twice as long as the oven residence cycle or, lastly, in the case of molding vinyl plastisol, the 4-arm machine may be provided with double ovens so that each oven may halve the residence time.

To help yield higher quality parts, the fixed-turret machines have been equipped with inert gas mechanisms which provide low air pressure during cooling to prevent shrinkage of the part, or provide nitrogen during the oven cycle to purge oxygen from the mold, a process which reduces oxidation and aids in the physical properties of parts. Reverse arm rotation in the oven enables the powder to flow better into more difficult crevices of the mold.

9. THE INDEPENDENT-ARM MACHINE

The independent-arm machine came in to prominence in the late in the 1980's and is the common style purchase in the rotomolding industry today. The Independent-Indexing Carriage Style machines provide the versatility of shuttle machines and the fixed-arm turret machine, but they provide a more automated and sophisticated means of handling the process cycles than any combination of other machines.

The independent-indexing carriage machine uses a forced-air recirculating oven and forced air and water cooling chamber, and is uniquely provided with five work stations. Hence, the molder can set the machine up to provide a molding process that suits his needs, be it heating – double ovens; cooling with double coolers; or loading/unloading with single or double cooling stations. The independent-carriage machine can be provided initially with one carriage, and at the molder's option he can increase to a second, third, and up to a fourth arm and carriage to provide more automation and high productivity. Figure 22 illustrates an operator sequence from an independent-indexing carriage machine.

The machine consists of a self propelled carriage to index each arm around a center post. A horizontal arm is mounted on the carriage to carry the mold. In effect, arm and carriage assembly is its own "independent machine". Each of these four "independent machines" are mounted on one common track assembly, and share a common oven and cooling chamber, and a common center post.

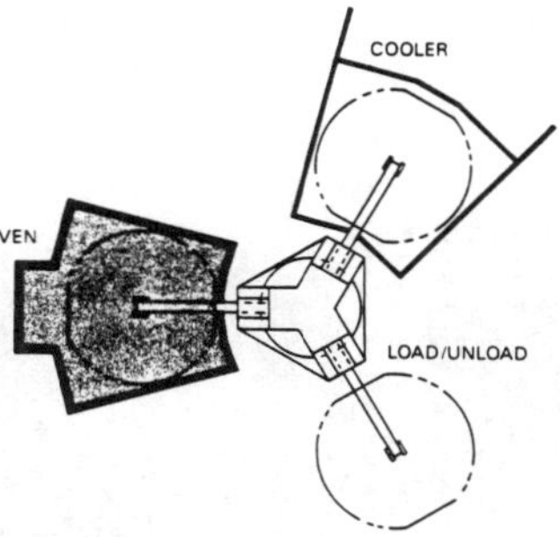

FIGURE 17 3-station, fixed-arm turret machine

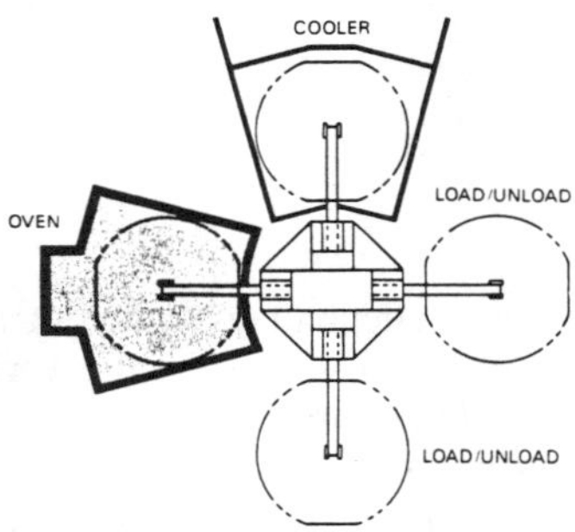

FIGURE 18

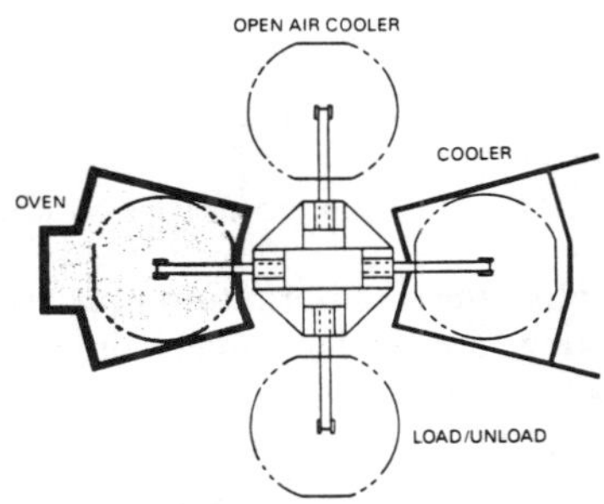

FIGURE 19

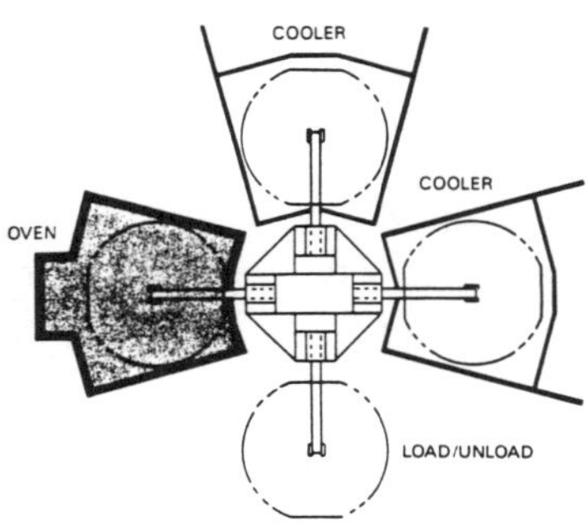

FIGURE 20

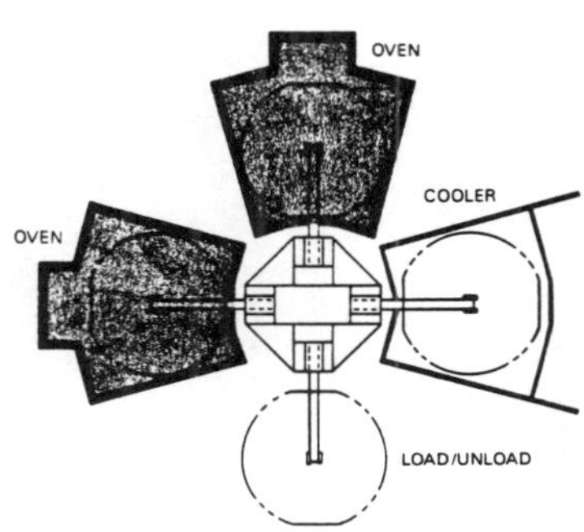

FIGURE 21

FIGURE 22 Independent-indexing carriage machine

Unlike the fixed-arm turret machine, one arm does not depend upon the other for indexing, and in fact, many times you will find dissimilar cycles on the various arms of the machine. Each carriage can have its own dwell times in the oven, cooling, and loading stations as required to produce an optimum part. Figures 23 thru 27 illustrate a typical cycle of a 4-arm machine and how its independent-carriage movement can provide advantages to the user.

The versatile independent-indexing carriage machines offer the advantages of shuttle machines in that each carriage can carry a very high load, enabling it to produce large parts.

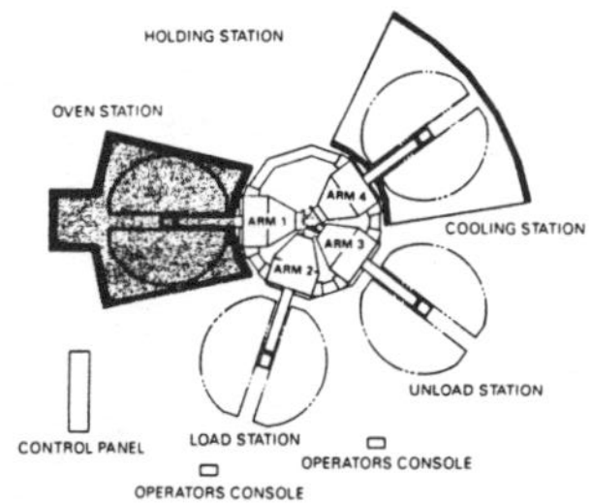

ARM 1 has completed the oven cycle, but ARM 2 loading operation is not complete.

FIGURE 23

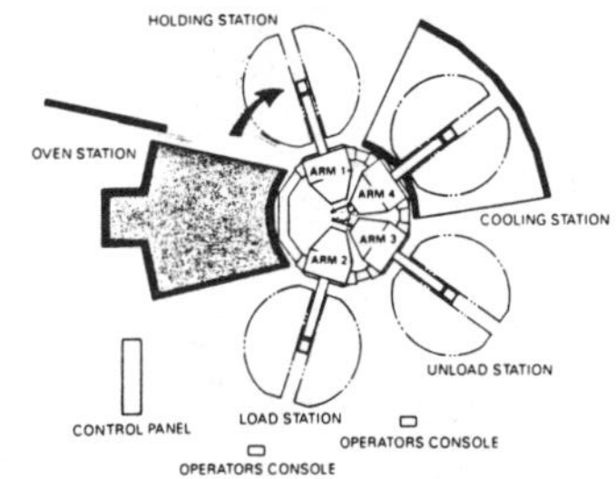

Oven exit door opens, ARM 1 automatically indexes to the holding station continuing its rotation. **Part does not over cure.** Exit door closes. Cooling begins in holding station.

FIGURE 24

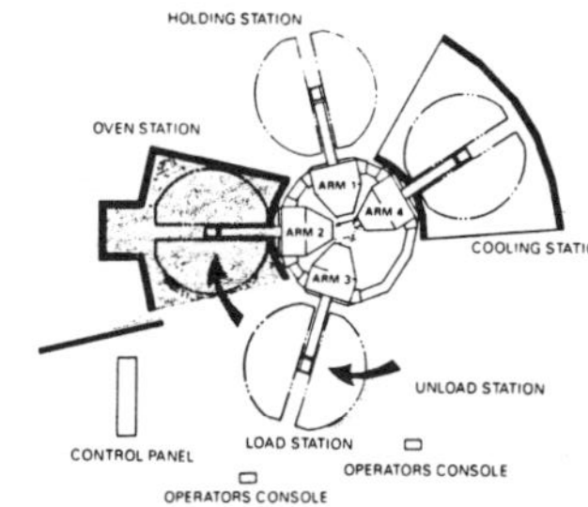

When ARM2 loading is complete, oven front door opens, ARM 2 indexes to oven and begins cycle. If ready, ARM 3 indexes to load station.

FIGURE 25

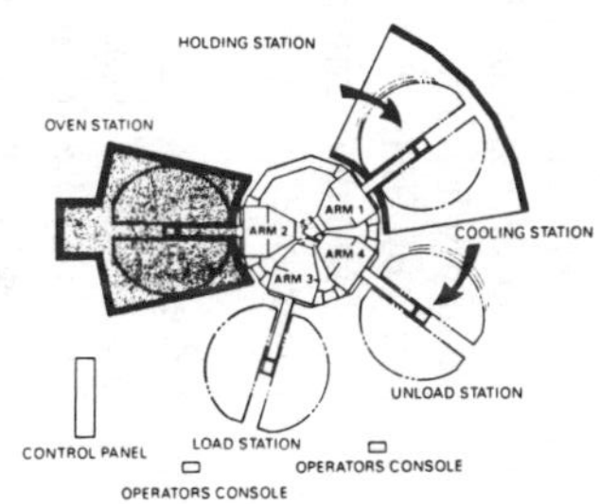

If cooling is completed, ARM 4 indexes to unload station. ARM 1 will then go into cooling station for cool cycle.

FIGURE 26

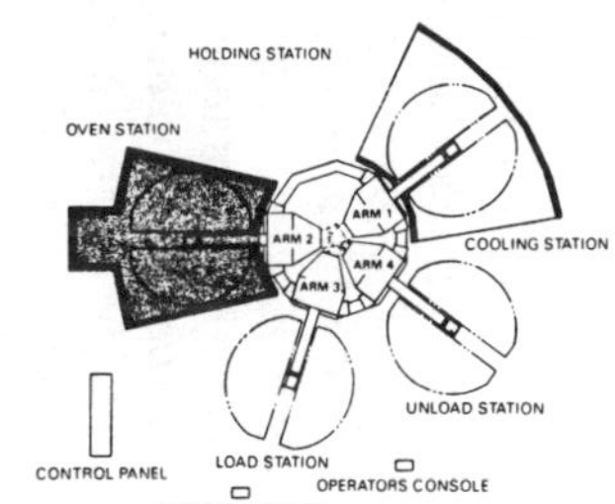

Machine is ready to repeat automatic sequence.

FIGURE 27

CHAPTER 5
Sheet Metal Moulds

T.J. Taylor

1. INTRODUCTION

Sheet Metal Mould making, in practical engineering terms, presents no specific technical difficulties because cold forming, welding and dressing are all part of standard sheet metal practice. Unfortunately, the intrusion of other factors tends to complicate the outcome rather than the means by which the end result is achieved. With most specialised engineering applications, practical and technical expertise has to be combined with a detailed knowledge of the end process, to enable the supplier to qualify as a toolmaker. Of course, this is not unique to Rotational Moulding. The same is true of toolmaking for Injection Moulding, Blow Moulding, Compression and Transfer Moulding.

For the Designer the parameters are much simpler:

Know how; Know why; Know who.

To establish the 'know how' it is necessary to look simply at the three most widely used forms of manufacture for moulds in the Rotational Moulding process. Fabrication, Cast Aluminium, Electroforms are the techniques, but what factors influence the choice of one of them? In particular, in the context of this chapter what leads to the selection of fabricated sheet steel as a means of producing a production tool?

The three tests to apply are

A. Complexity of Form
B. Quantity of Moulds required
C. The Appearance of the Product surface.

A. Complexity of Form

The selection process begins with an overall impression of physical size of the product, which alone may be sufficient to eliminate some of the methods of tool manufacture. From size the progression is towards the general part geometry. How complex are the component shapes? How many changes of section are there? The key factor is to identify what is practically possible. Is it a form which can be readily translated to a layflat image or developed by a machine tool or simply converted to a full size model?

The final factor in regard to complexity is to assess what type of product is to be produced, what is its function, where will it be used and what material is it to be moulded in?

Having gone through this critical appraisal of the product, its design and its function, it should now be clear which of the processes is most appropriate for the manufacture of the mould. However, two other factors must be considered to confirm that the correct selection has been made.

B. Quantity of Moulds

If it is necessary to have more than one mould in order to increase production rates and/or reduce item cost, then consideration must be given to the need for uniformity of the repeat units. Although some sheet metal toolmakers would disagree, it is generally more difficult to achieve repeated uniformity with metal fabrication compared with the other two methods. Also, there is a cost saving in making repeat cast or electroformed moulds from a single pattern compared with fabricating each mould individually from sheet metal.

C. The Appearance of the Product (or surface detail)

This is the final piece of information required to enable the correct mould making method to be chosen. Although the surface of sheet metal moulds can be highly polished and can be chemically etched, the range and degree of surface finish are limited. There are few such restrictions with Cast Aluminium or Electroform Moulds, where the desired detail is readily applied to the original model.

The inter-relationship of these three factors is illustrated in Fig. 1. The three intersecting circles represent the three Single Identities:

A. Complexity
B. Quantity
C. Appearance.

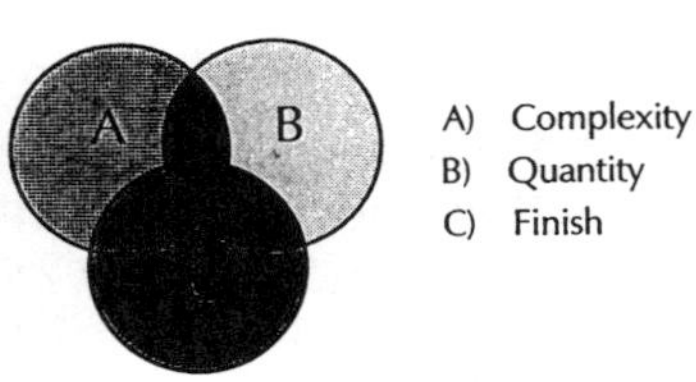

FIGURE 1

If identity A dominates then this will almost certainly give rise to the production of a fabricated mould. However, increasing importance of identities B and C in the product make the decision less clear-cut. The areas of intersection represent the change of emphasis from Fabrication to Cast Aluminium to Electroform with the central matrix excluding Fabrication completely.

1.1. Cost

So far one important consideration has been omitted from the discussion - Cost! Although the actual cost will be generated in the form of a quotation by the individual Toolmaker, it is as well to understand how this cost is devised. This represents the 'know why' factor for the designer. Suppose that the decision has been made that the mould is to be produced from Mild Steel Sheet by Fabrication. First of all, what information will need to be passed to the Toolmaker and then how will he determine his costs?

It should be emphasised at this stage that as much information as possible regarding production requirements should be conveyed to the Toolmaker but basically he must know:

1. The moulding material.
2. The tool construction requirement of the moulder (mould material thickness).
3. The position of the Parting Line.
4. The number of tools (or impressions).
5. Required surface finish.

With the component drawing and with a knowledge of the moulder's requirements, the toolmaker can begin to assess the number of man/hours required for fabrication. The process begins by forming a plan of construction that will take into account the following key areas.

Parting Line

What is its position and how complex is it?

Can it be produced in one plane or are there areas, above or below the general plane, which for the sake of part withdrawal, need to be at a point of parting?

Is it necessary to have more than one parting face and are there any special features through which the parting line must pass?

Each additional complication or deviation from a flat, single, plane parting face will accumulate extra hours.

Method of Construction

This evolves from the basic component geometry, and how complex the shape is. The toolmaker must decide within the capabilities of plant and workforce whether component parts can be produced in-house or need to be subcontracted. Items such as stock spinnings, or perhaps pressings, derived from a simple press tool can be purchased. This will be done to keep the amount of hand work to a minimum. Much of this part of the calculation will be affected by what gauge of steel is to be used for the mould. If this is not specified by the moulder, the toolmaker will probably have a standard which is best suited to his own normal methods of construction.

Inserts

In assessing the nature and position of the parting line the need for, and location of, inserts may have already been taken into account. However, in many cases it is also a requirement to have localised removable areas for ease of part withdrawal. Such is the case with threaded bosses or holes. The work can be further complicated by a series of such inserts which may require a method or arrangement to compensate for the effects of part shrinkage during cooling.

Engraved areas are normally dealt with as fixed inserts. The graphic detail is usually subcontracted to a specialist engraver, and the engraved detail, in the form of a panel, is let in to the surface of the tool as required.

Special Features

There may be a need to either exclude heat from areas of the mould or to induce extra heat in some zones. The former is usually the case where a large open area is required, and so to reduce waste it is desirable to minimise the amount of material deposited in this part of the mould. In the latter case the designer may have identified areas of the mould where the product wall thickness is to be increased. Alternatively, the toolmaker may have identified a problem with the shape of the product and feels it is necessary to induce heat in some areas to sustain uniform wall thickness.

The generalised term, special features, however, can include any element of mould construction that is intended to aid the moulder.

Clamps and Clamping

The type and method of clamping may be a pre-condition laid down by the moulder, but since this forms a vital part of construction it is necessary to know the number and location of the clamps to be utilised. These factors can have a significant effect on the cost of the mould.

Finish

Last, but by no means least, is the type of finish required on the moulding surface. To achieve a quality finish, to whatever degree, involves the toolmaker in making the right decisions concerning materials, welding and forming methods. This particular aspect of sheet metal mould making, or more specifically the work involved in order to achieve this end, consumes by far the greatest percentage of time allocated to the job (perhaps as much as 50%). The moulder/designer is urged, therefore, to consider carefully the importance of the specified surface and not to take this decision lightly.

Summary

Once the above factors have been quantified, it is then simple arithmetic to arrive at an estimate of the mould cost. The number of hours allocated for each of the key areas are added together and then multiplied by the Labour and Overhead rates. The Material content cost is then added and the Factory cost is established.

The following diagram (Fig. 2) illustrates a typical Cost Estimate sheet.

LABOUR COST		TYPICAL SUMMARY	
1	Parting Line/s	20 hours	
2	Method of Construction	120 hours	
3	Inserts & Cores	40 hours	
4	Special Features	10 hours	
5	Clamping	5 hours	
6	Finish	100 hours	
	TOTAL HOURS	295 x O/HEAD	£______
		x LABOUR RATE	£______
	MATERIAL CONTENT		
1	General Construction		£______
2	Machined Items		£______
3	Proprietary Items		£______
	TOTAL FACTORY COST		£______

FIGURE 2

2. TOOL CONSTRUCTION AND DESIGN

In the evaluation of mould cost, the framework of the mould making process has been outlined. It is now necessary to put substance to the framework and consider the subject of mould construction in more detail. In the following sections, the process of mould construction is described to assist the moulder/designer in identifying the constraints under which the toolmaker works and which inevitably influence the quality and cost of the mould.

2.1 Parting Line

Since the construction, operation and possible success of the moulded product evolve from the parting line it is fitting that it is considered first in relation to mould design and construction. It is obviously preferable to be able to say 'The parting line can be here, because that is where it is intended to be' than for the position to be decided as a matter of convenience or compromise. No matter how well the mould is made, there will be a witness line present. Whether or not this is to be masked or disguised needs to be resolved at the outset because it can lead to production problems or, worse, product failure. It is important to recognise the crucial importance of the parting line. It is the area around which the whole mould is fabricated and it is the area which suffers the most use and abuse. To the toolmaker, more often than not, it is more than this. To him it is the strength of the mould, it is his gauge and it is his reference datum. In order to understand this, it is first necessary to look at the various types of rims or flanges which contribute to the construction of a mould parting line. The most common types are illustrated in Fig. 3. They are:

(a) Flat bar flange
(b) Angle flange

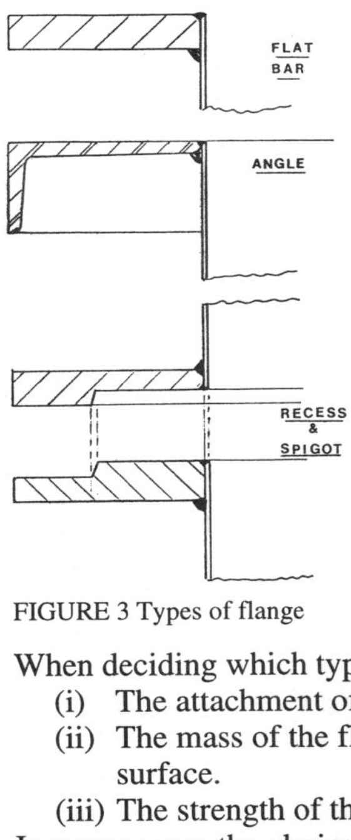

FIGURE 3 Types of flange

When deciding which type to use it is necessary to consider:

(i) The attachment of the parting line to the mould surface.
(ii) The mass of the flange and its effect on temperature at the mould surface.
(iii) The strength of the parting line.

In many cases the choice is a matter of personal preference by the mould maker or moulder but there are merits and limitations of each type. The flat bar gives a greater attachment area to the mould but also can give rise to heat starvation due to mass, particularly with thin wall mouldings or heat sensitive materials. However, it lends itself better to the adoption of a stepped profile, often used for location with simple geometric shapes, or as a venting baffle. The angle section flange has a better strength to weight ratio and reduces the heat starvation problem, but has a lesser attachment area and is generally limited to dowel or side plate location.

In both cases, however, precise mating of faces is crucial. At the beginning of construction, the sheet metal mouldmaker will invariably begin by producing a pair of matched flanges (as shown in Figs. 4 and 5) where both inside and outside edges are in line. This pairing serves as a form gauge and reference throughout construction. Knowing the exact relationship of inside and outside faces allows the toolmaker to determine any misalignment or mismatch once the mould is together and the view inside is denied him.

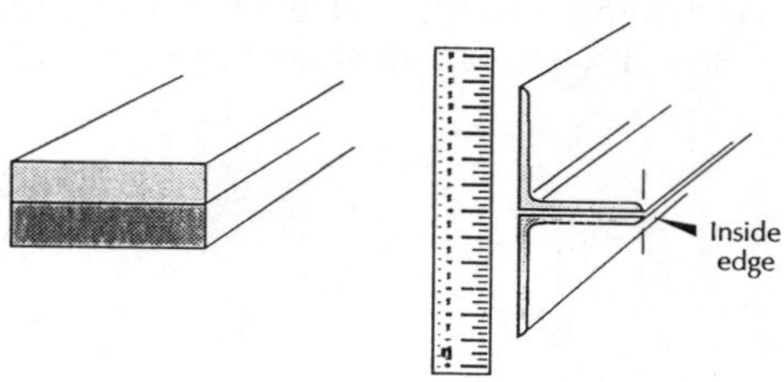

FIGURE 4 Mating flanges

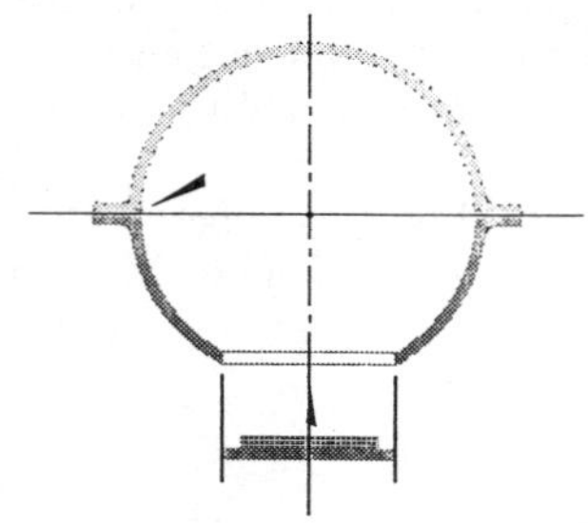

FIGURE 5 Precise match of two mould halves

In some instances, however, where precise matching is particularly important, a panel can be omitted until both inside faces are aligned, or on larger moulds it is possible to work totally inside the closed mould to achieve the desired result.

It is important to understand how the flange is assembled to the main body of the mould in case future work is needed to be carried out. In a typical construction using an angle section, there are two welds at the point of contact, one at the top which acts as a sealing weld as well as forming the mould edge, and one underneath which is the main attachment weld (see Fig. 6). In some constructions this attachment weld is not continuous but intermittent. Therefore, if for any reason the mould edge is removed, or more importantly the sealing weld is removed, the joint is exposed. This not only weakens the flange attachment but may also cause blowholes to appear in the moulding.

If, for example, the mould edge is required to be chamfered or to be any shape other than square, the initial construction should be adjusted to accommodate.

This example is quoted to demonstrate, firstly, how a simple modification, without thought, can impair the mould performance and secondly, how important it is to convey such detail to the toolmaker before construction begins.

2.2 Method of Construction

The basis for good construction has its foundation in good article design and drafting practices in conveying correct and relevant information to the mould maker. Whatever the form to be fabricated, it is inevitable that the toolmaker will have to consider which elements can be machine formed. To do this he will be mindful of reducing the amount of welding in order to minimise the effect of distortion and

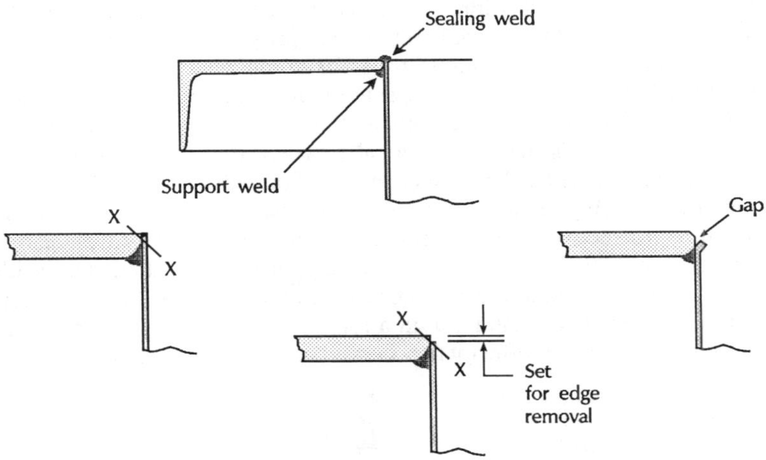

FIGURE 6 Removal of mould edge

thus keep the final operation of finishing and polishing simple and less time consuming.

However, at some stage the component will have to be developed or unfolded onto a single flat sheet of prime quality cold rolled steel as illustrated in Fig. 7. Obviously where parts can be machined or machine formed, by using any of the standard engineering practices, advantage is gained. This is because not only are

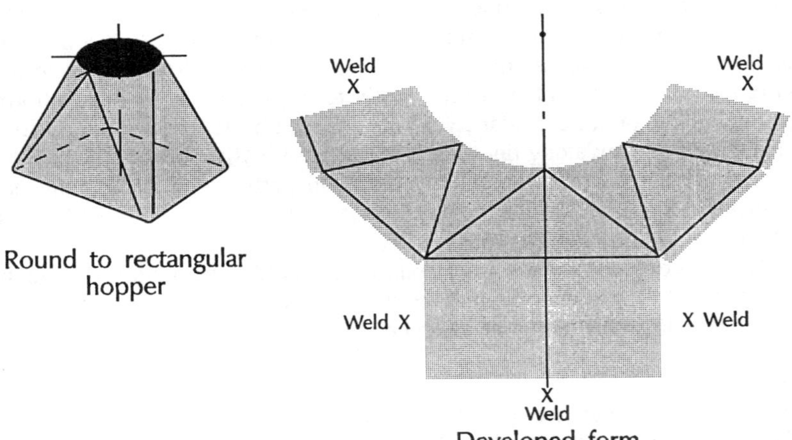

FIGURE 7

the true geometric shapes formed (and perhaps to a greater accuracy) but also the amount of welding is reduced. It is imperative that welding skills are of a high standard, but endeavouring to reduce the amount of weld produces enhanced benefits.

The most common welding technique used for component fabrication is TIG (Tungsten Inert Gas). This enables the welder to obtain maximum weld penetration with minimum heat input. In practice a combination of techniques is used depending upon the technical requirement but generally, where joining is required about the mould surface, TIG provides the most economical process.

Joining metal by welding creates forces along the weld seam and this results in distortion. In order to recover the contractual loss it is preferable to planish (or beat) the weld as shown in Fig.8. This serves a dual purpose. It flattens the welded area along the line of weld and it compensates for shrinkage. Where welding occurs through the line of a radius, planishing with a form tool reproduces a more acceptable curve adding to the overall effect when viewed at an edge or styling line.

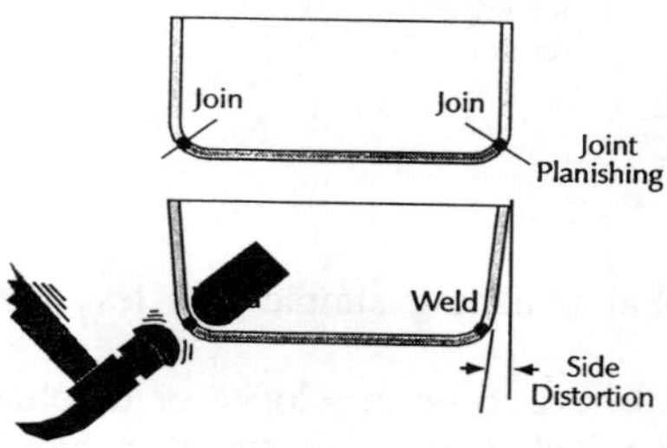

FIGURE 8

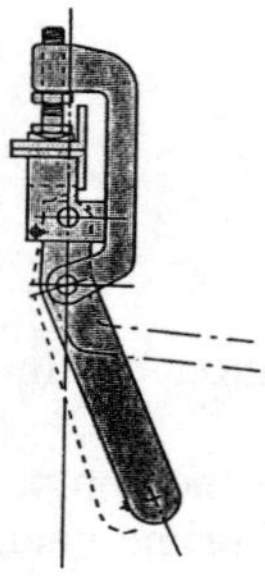

FIGURE 9

When assembling a series of preformed component parts, which individually may be semi-finished, it is good practice to tack-weld (a small localised weld) the assembly. This allows the overall form to be aligned and viewed before completing the joints. Even at this stage careful thought as to how the final welding is to take place can reduce the amount of distortion likely to occur. Instead of concentrating on one particular side or area, a balance can be achieved by alternating the position. This reduces heat build-up along lines likely to cause excessive stress.

With component parts of the mould formed and partly assembled, there now must be consideration for attachment of these parts to the preformed mould flanges moving towards total mould build-up.

This will then conclude the technical phase of mould construction. It remains only to complete all the weldings so that the two halves (or all parting faces) can be bedded together to form the complete closed mould with a secure parting surface.

2.3 Clamping

With two level, equally matched, parting surfaces achieved, it now becomes necessary to maintain the integrity of the parting line by adequate and secure clamping. The selection of the method or type of clamping is usually made by the moulder and will be directed to what is best in his moulding unit. This is likely to range from single nut and bolt fixing to multiple frame clamping. The most commonly used method, however, is a form of reverse action toggle clamp as illustrated in Fig. 9.

Whatever the means of clamping, the object is always the same, i.e. to provide a secure mould parting surface. The speed of demoulding, and therefore the production cycle, will determine to some extent the eventual method of effecting a closure.

Three major considerations for the toolmaker, should be:

(i) To provide the clamp force as near as possible to the moulding edge (see Fig. 10).

(ii) The number and position of the clamps.

(iii) The on-going effectiveness of the clamp (maintenance).

One of the commonest causes of deterioration of the mould quality at the parting line is badly maintained clamping.

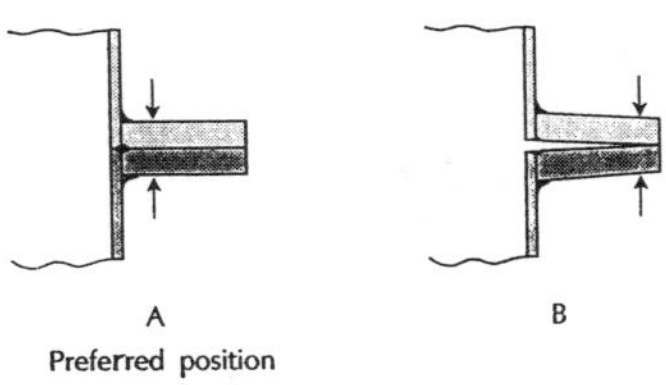

FIGURE 10 Clamping

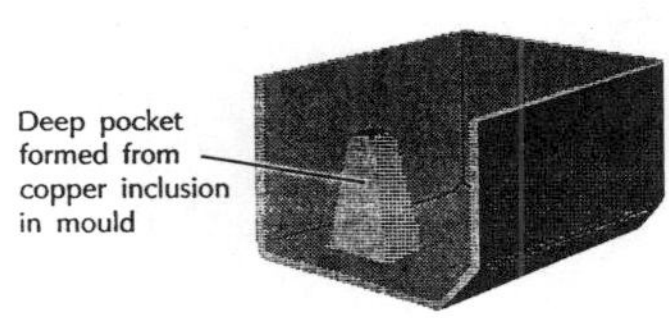

FIGURE 11

2.4 Inserts and Cores

It is difficult to be specific about inserts since they can be of such a variety. The only features that all have in common are that they should be so constructed to effect easy removal of the component and to attract sufficient heat to obtain the desired material thickness.

Generally they can be classified into 4 types:

(a) Those that are FIXED.
(b) Those that are INTERCHANGEABLE (Semi-Permanent).
(c) Those that are REMOVABLE.
(d) Those that are MOULDED-IN.

(A) Fixed Inserts

These could be considered as part of the general tool construction, but more specifically they are treated as an insert because of some special feature. For example, a deep pocket as shown in Fig. 11 is perceived to be an area over which material deposition may be difficult to achieve due to a heat differential. In this case the insert may be constructed from:

(i) a lighter gauge material or
(ii) a more heat conductive material.

Other examples of fixed inserts are:

(a) 'Kiss points': areas of the mould which are raised to form a narrow gap with adjacent or parallel surfaces (see special features).
(b) 'Engraved Logos or Instructions': preformed on a separate piece of material, as a machine function, and added to the mould at a later stage.
(c) Machined components: Areas of precise or specific detail which can be produced accurately by using a standard engineering facility (Fig. 12).

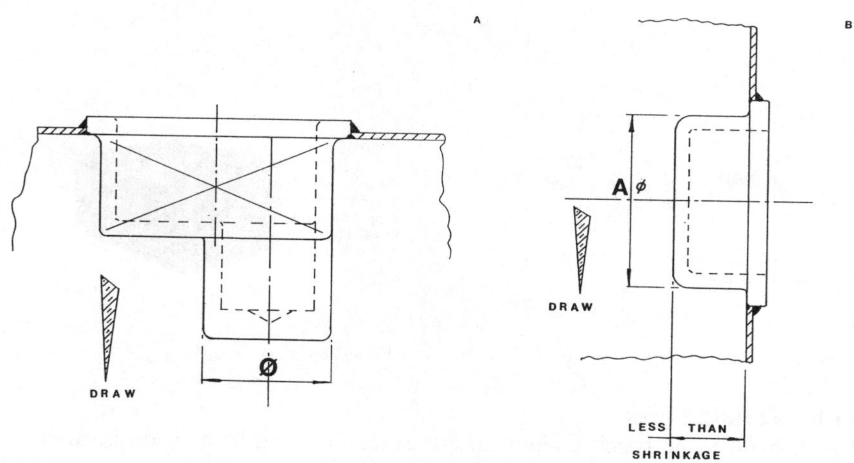

FIGURE 12 Fixed inserts

(B) Interchangeable Inserts

These are basically simple inserts usually in the line of draw which are an 'either-or' situation for the component part. A common example is where a standard product is used for more than one customer. The individual customer's identification can be inserted into a moulding face and changed to a blank plate or an alternative identification as the component source changes.

Since the production requirement is usually not one of constant removal these can have a more semi-permanent nature as opposed to the quick release method of clamping (see Fig. 13).

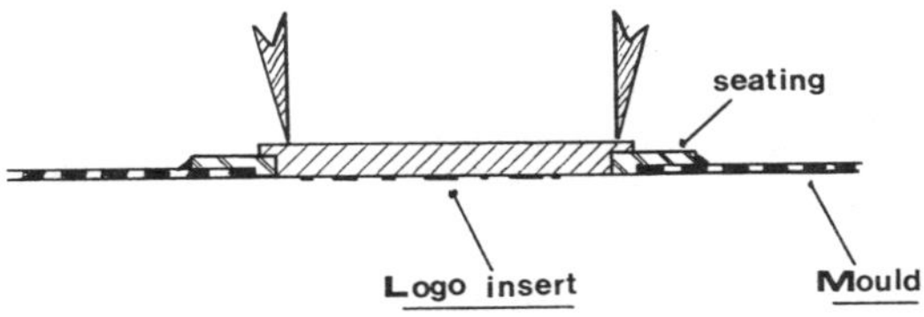

FIGURE 13 Clamp

(C) Removable Inserts

These are by far the most complicated and diverse range of inserts and before considering some examples it is well worth emphasising that regard must be given to the effect of continual use. The insert should be protected from the use of undue force on removal. Any clamping should be substantial and easy to maintain (see Fig. 14(a)).

The reason for an insert to be removed is to effect de-moulding but what it actually forms is the subject of most diversity. A few of the most common uses are as follows:

Removable Threads

Most standard engineering threads of finer pitches and small nominal diameters do not form well in rotational moulding (in some cases it is preferable to form a hole or boss which is then threaded by hand tools). However, modified thread forms can be moulded-in and where this is a requirement the need is for a removable insert, premachined, to provide all the necessary requirements. An example is illustrated in Fig. 14(b).

The provision of such an insert is a simple engineering operation but to comply with other mould manufacturing criteria it often needs a little more thought. For example:

(a) How is the insert to be held in place - not just once but for every production cycle?

(b) What sort of location and seating is to be used to give continual trouble-free bedding? If the latter cannot be guaranteed for the life of the tool, can replacement components be installed easily?

(c) Finally, is the effect of shrinkage likely to cause problems such as:

  (i) Incorrect pitching.

  (ii) Thread shear or

  (iii) Distortion.

These types of problem arise for the following reasons:

(i) Incorrect pitching: Thread pitch correction is not normally required for rotationally moulded threads as they are generally of short length with adequate fit tolerance. If it has to be catered for, however, it is handled in the same manner as for injection moulded components - the thread is machined with dimensions which allow for material shrinkage.

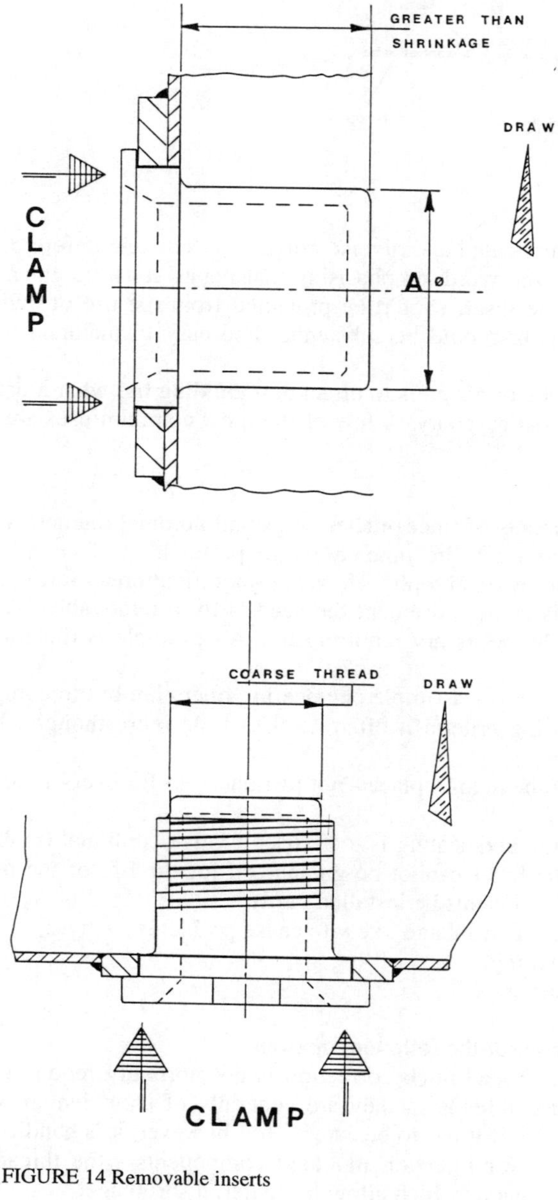

FIGURE 14 Removable inserts

(ii) Thread shear: This can be a problem where the thread has to resist shrinkage pull due to the component being more rigidly held in other areas and the thread strength is insufficient to resist shear. A method of dealing with this is illustrated in Fig. 15.

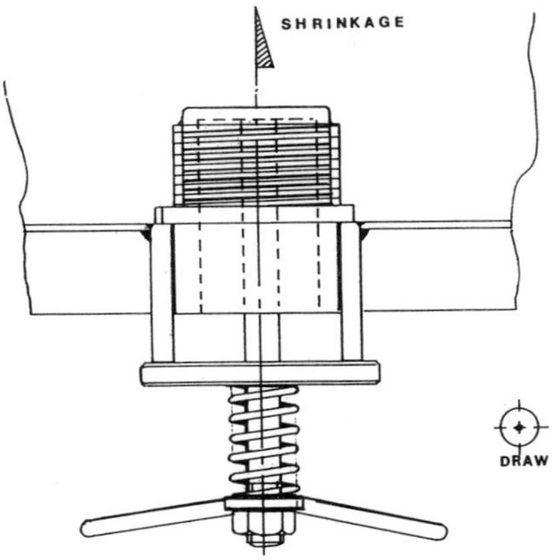

FIGURE 15

(iii) Distortion: This is similar in some ways to thread shear but here the shrinkage force is not axial to the thread diameter. Where thread shear or distortion has to be taken into account, the construction can be such as to allow the insert to move during contraction. The insert, however, must be firmly held in place during the moulding cycle. Where springs are used for this purpose (Fig. 16) it is important that the moulder appreciates the importance of regular checks for uniform and effective spring tension.

Removable Handles

A familiar requirement with products such as drums, hoppers and containers is for some form of lifting handle. Invariably in the design, they will be a restriction to part removal and therefore will require the use of a removable insert. A particularly useful solution is shown in Fig. 17. Here, use is made of the dissimilar expansion of two metals to obtain a secure parting face at one end where clamping has no effect. The handle used is a casting (phosphor bronze or aluminium) and the bedding surface is an insert of copper.

Moulded-in Holes

The use of copper, brass or aluminium is of particular advantage when wishing to produce 'through holes' against the line of draw. Where the component details are too small to allow heat to be applied along the inside surface, a more conductive material will allow a transfer of heat from the main mould surface to take place (see Fig. 18). The inserts can either be removed before demoulding or after the moulding has been extracted depending on the method of fixing.

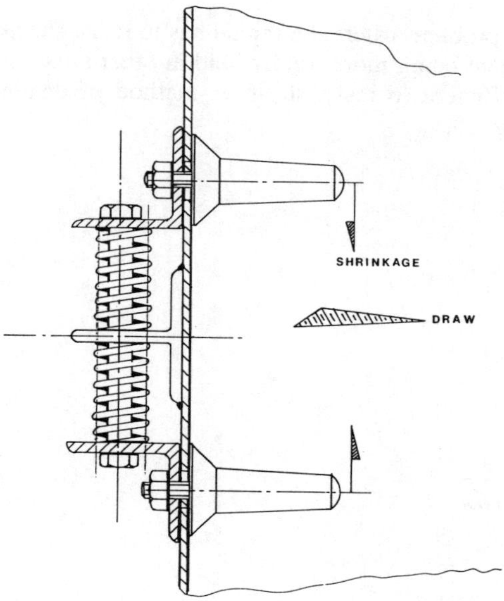

FIGURE 16

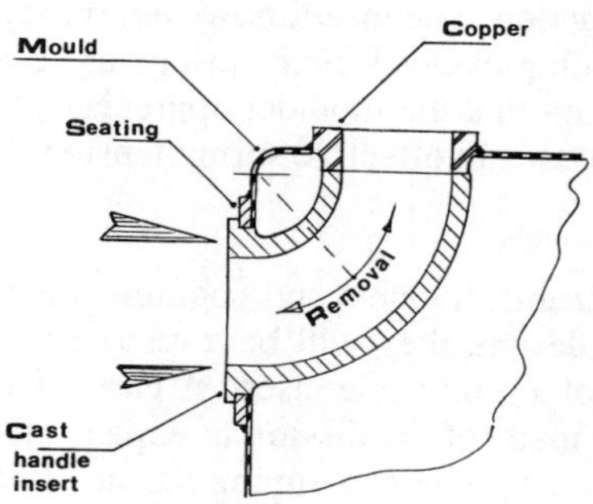

FIGURE 17

(D) Moulded-In Inserts

With this type of insert, the mould and mouldings criteria are slightly different. Here the insert becomes part of the finished component and is replaced within the mould after each production cycle. Nevertheless they still have to perform similar functions. They must attract sufficient heat from the outer surface of the mould to allow material to be deposited and they must allow (and not restrict) material flow during moulding.

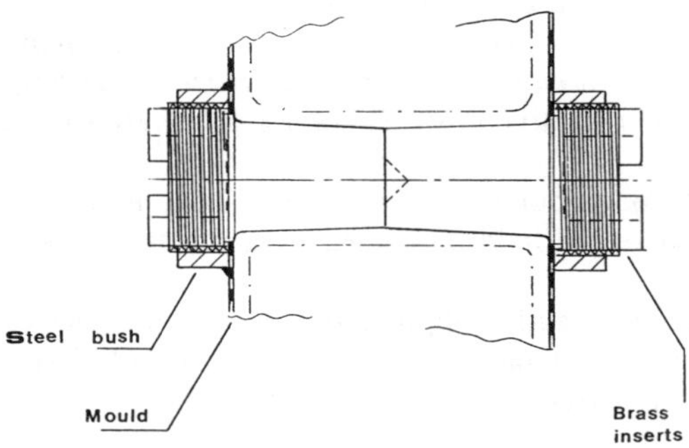

FIGURE 18

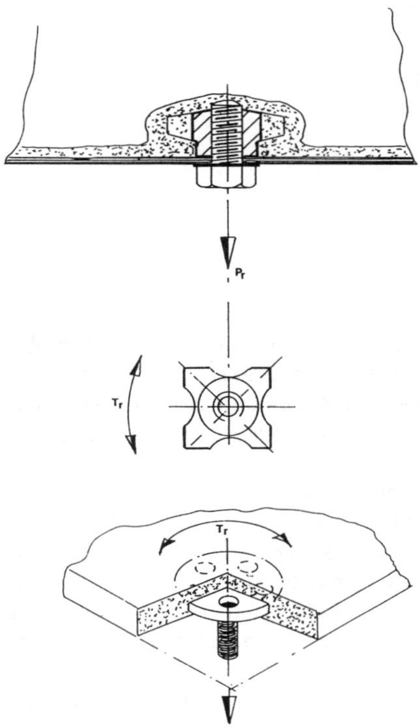

FIGURE 19 Moulded-in inserts

In addition to these production requirements, they may also be required to resist torque, or tension, or both, after moulding, as the product performs its function. Some examples of this type of insert are shown in Fig. 19 and are mainly of a threaded nature. The grouping of this type of insert should be treated with care. Where pitching or precise placement is called for on the moulded component, regard must be given to the effect of shrinkage. It may be necessary to contain the group of inserts as a whole so that their relative position is unchanged after moulding.

Another method of producing 'through holes' against the line of draw is with a 'moulded-in' insert (see Fig. 20). Here the hole is formed by a metal tube which stays with the moulding.

There are a continuing number of specific applications where the use of a removable insert is called for. However ingenious or complicated the solution, the underlying principles laid down at the outset of this section always apply.

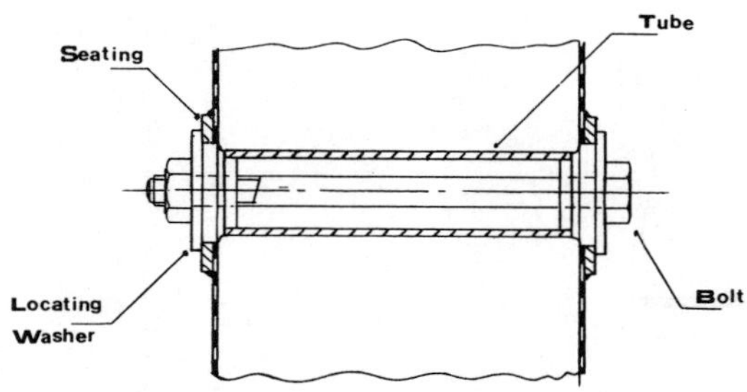

FIGURE 20

3. SURFACE FINISH

Probably the most labour intensive part of tool fabrication is that required to obtain a quality finish. It is important therefore that all the previous work in assembling the tool has been correctly carried out. Good practices and care in construction enable the toolmaker to achieve the desired finish more readily, whether it be a matt surface or mirror polish.

By the time finishing takes place, all welding is complete and all distortion has been removed. The mould is complete except for the feature which determines the quality of the product. There is no easy way to obtain the end result. It is ironic that the better the finish produced, the more apparent become the mistakes or bad practices which were left unattended in the early stages. The preparatory work whether for mirror polish or somewhere below is the same; the difference is only determined by where the work ceases.

Mild steel is a relatively soft material and is therefore susceptible to scratching. The toolmaker will systematically work those areas which have received previous attention (welding or forming) with an abrasive, gradually reducing the grit size until the desired effect is obtained. The real object is to provide an even overall smooth surface. Plain areas (where no previous work has been carried out) can be left until the later stages. There is little point in producing scratches on an unworked surface in order to remove them later.

Since surface finish is 'HAND WORK', it can still be a source of problems if care is not taken. The careless use of power tools with abrasive heads can weaken a surface. Over-enthusiastic use of polishing equipment, to get a quick result, can generate localised heat, thus producing distortion. In tools which are being surface treated using a 'Grit blast' process, excessive blast pressure or dwelling in one area for too long with the blast head can also induce metal movement.

The polishing process gives a limited range of surface finishes to steel tools. It is possible to obtain textured surfaces but within strict limitations. Standard steel sheets are obtainable with a patterned surface and these may be folded or formed. The obvious disadvantage in this case is that, where joining or welding takes place, the pattern will be removed by dressing or beating.

A patterned surface can be applied by the process of electro-chemical etching but, again, a welded seam can mar the end result. No matter how well the weld has been flattened or dressed in the preparatory stages, the line is highlighted and is visible as a change in pattern intensity. In any event, this process is expensive to perform and may rule out the choice of FABRICATION in favour of one of the other methods, if a surface pattern is called for.

4. SPECIAL FEATURES

4.1. Heat Inducement

One advantage of fabricated moulds is that areas of dissimilar metals can be 'built in'. This is particularly useful where it is required to induce heat more readily to the moulding surface. These areas are not easy to identify but generally can be said to be:

(a) areas which encounter heat starvation, due to the mould construction, or

(b) areas where, due to component geometry, the moulding material will be forced to pass over more quickly. By inducing heat quickly into these areas, the moulding can be allowed to attain the thickness expected to be produced over the remainder of the surface.

Copper with its excellent thermal conductivity is the normal insert material but should only be added to form small areas. The addition of a large area of copper is likely to cause distortion or buckling due to differential expansion. If the area in question is likely to be large, the copper can be inserted in the form of strips. In some cases it may well be that the desired effect can be obtained by using a lighter gauge of steel, but this is limited and would result only in a small increase in thickness over a large area and would have no effect if applied to a small area.

The insertion of a dissimilar metal is only one solution. Two other methods are widely used:

(i) To construct on the outside of the mould a feature which will produce a faster air flow over the surface. Deep pockets or ribs in moulds can provide 'dead spots' where little or no movement of air is experienced or where surrounding convection currents exclude the hot air (see Fig. 21).

(ii) The alternative method, which is well practised in cast moulds, is to use a Heat Pipe. This is a proprietary item. It is inserted into the mould, often in groups, and effects rapid transfer of heat by means of an enclosed liquid which is vaporised and then condensed producing latent heat which is given up at the mould surface. A typical application is shown in Fig. 22.

(iii) A rarely used but effective method of overcoming mould 'dead areas' is by way of a forced air supply. Compressed air is fed to a storage chamber which is heated. Then through a series of rotating joints this air is directed into the constricted area. This is a complicated 'Engineering' solution but most effective if there is the likelihood of the production requirement continuing for some time. It is illustrated in Fig. 23.

(iv) A more recent introduction which overcomes many of the engineering problems associated with (iii) above is to introduce an 'Air Mover' to the problem area. Air Movers are compact units which require a compressed air feed but utilise the hot oven air and therefore do not require an air reservoir. They work by Venturi principle, drawing in the surrounding air and then expelling it at high velocity towards the 'heat-starved' target area. They can be used as a single amplifier or grouped into a multiple array for larger, difficult areas.

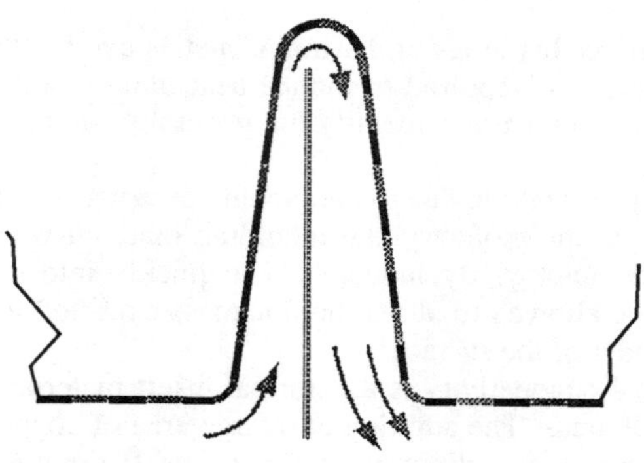

FIGURE 21 Heat compartment deep rib or pocket

4.2 Heat Exclusion

A more frequent requirement in a mould is to provide some form of insulated section to exclude heat. This is intended to reduce material wastage by masking areas where material is not required, eg open top containers. The degree to which material is excluded will obviously depend on the effectiveness of the insulation material. Tightly packed mineral wool to a thickness of between 25 and 30 mm is generally considered sufficient (see Fig. 24).

The practice of masking or insulating is often used by moulders as an answer to some of the problems set in the 'Heat Inducement' section. Where there is a

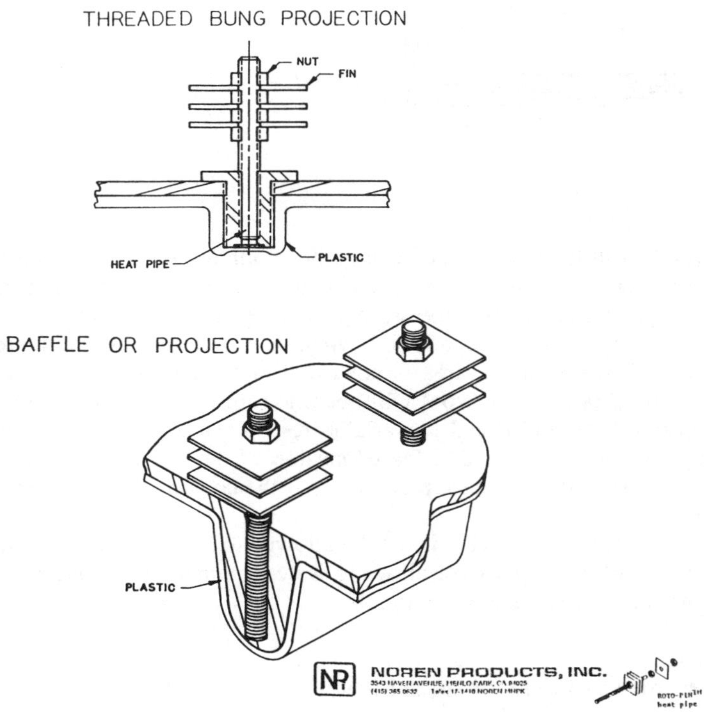

FIGURE 22

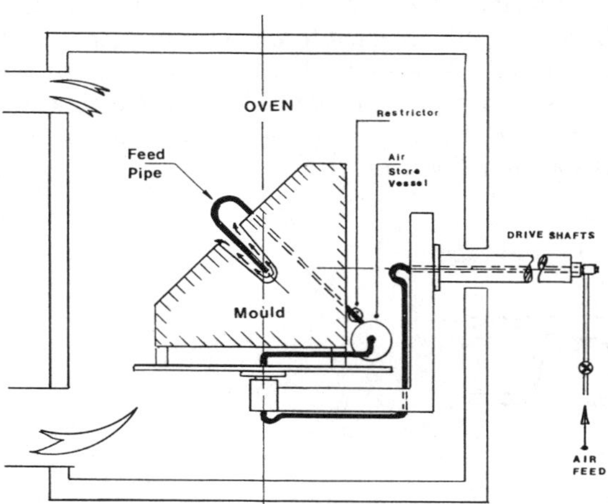

FIGURE 23

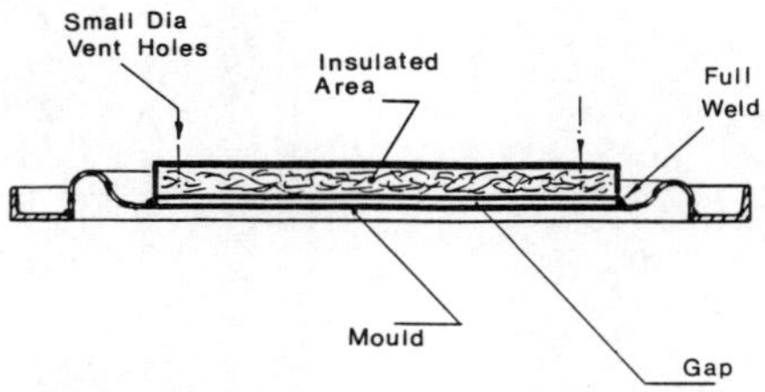

FIGURE 24

difference in overall wall thickness which cannot be countered by other means, the areas where the material thicknesses are greater are masked to slow down heat transfer and allow a better build-up in the areas previously starved of material.

An element of caution must be advised when constructing insulated areas. Mineral wools absorb water. If the area is not fully contained and water is used during mould cooling, the mineral wool will become saturated and the insulation efficiency increased. This could lead to variable results. However, if the area is totally contained there is a danger that the expansion of the air inside the cavity during heating will distort the surfaces, particularly if the constructional material is of a light gauge.

In practice a compromise is struck; air will be allowed to escape from the insulation chamber, but not at the expense of allowing water to enter freely and become trapped, thus degrading the surface of the mould.

4.3 'Kiss' Points

These can be regarded as inserts or special features. They are localised points on the mould where the mould wall is extended to almost touch a parallel or adjoining wall. The intention is to force the material to flow through a narrow gap and eventually to join the two surfaces together. These areas are used to enhance the strength of the product. There is a generally accepted formula for the gap required, and that is between 1.75 and 2.0 times the material thickness. In addition, the shape of the top of the kiss point insert is important. It should be constructed so as to let material flow through the gap during the heating cycle without bridging or joining prematurely. A typical design is shown in Fig. 25.

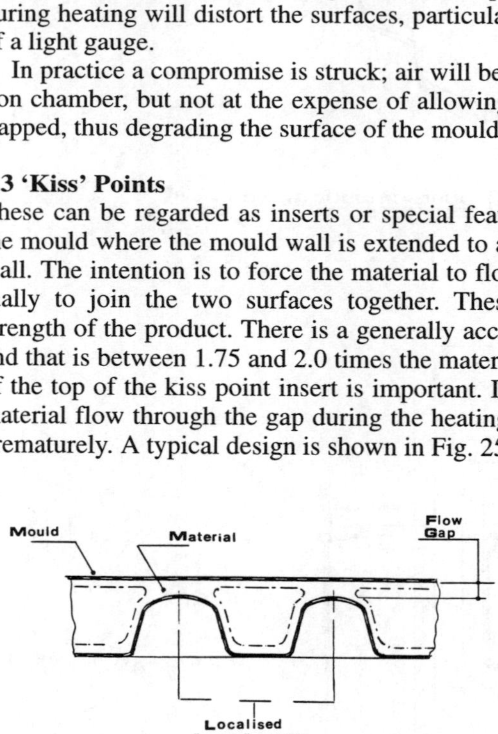

FIGURE 25

4.4 Charge Hoppers ('Drop Box')

In all aspects of tooling for a plastics process there are those special features which stretch the imagination and ingenuity of the tool designer. None more so than the use of 'Material reservoirs' or 'Charge Hoppers' for rotational moulding.

To generalise in simple terms, these are insulated hoppers for storing a charge of powder to be released at some stage of the manufacturing cycle. The key word is 'release'. Holding the required material is a simple task, but effecting a controlled, efficient release during the cycle is rather more exacting. There are several situations where it is necessary to utilise a Charge Hopper.

(i) If the volume of the mould is such that it is not able to accept the full charge of material, due to its high bulk density, then some means must be found to hold the extra material which can be released as the powder in the mould melts and adheres to the mould.

(ii) If an article is required to be moulded using two different powders. In this case the first material is charged into the mould and the second powder is stored and released once the primary charge has been processed.

The only difference between these two situations is that in the second case the material release is controlled mechanically while the release of material in the first case occurs as natural flow. In the first case the solution to the problem requires the use of an insulated reservoir as well as the provision of an area through which the material can flow. An example is shown in Fig. 26.

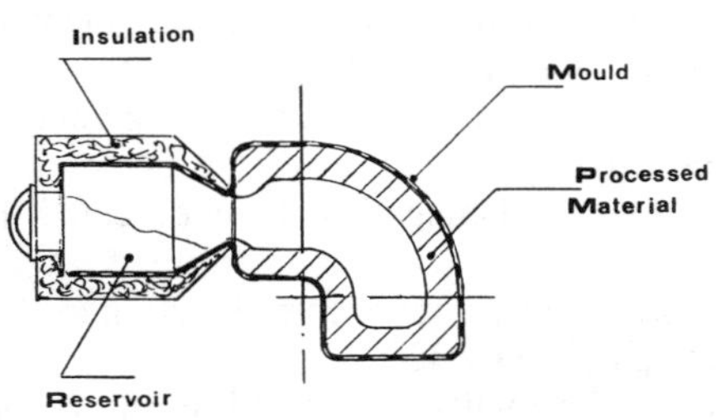

FIGURE 26

The introduction of a second plastic material also requires an insulated reservoir but in this case the reservoir is sealed until such time as a valve or gate is activated by some external means. An example of this is shown in Fig. 27. The design of the gate must be such that material must not be allowed to fuse over the gated (or valved) area, thus preventing the introduction of the new material.

5. VENTING

One of the most important aspects for the successful production of a rotomoulded product is venting, but so often it is put in as an after-thought or considered as a minor feature at the end of construction. Rarely is this necessary feature specified by the moulder. It is more often than not left to the discretion of the toolmaker. The

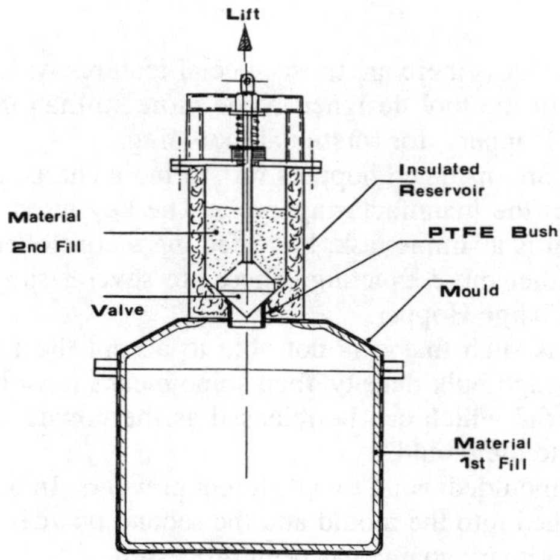

FIGURE 27

siting of the vent is often dictated by the component design, for example, a moulding area not normally visible in use, an area removed after moulding or an open end. All are useful sites for the positioning of a vent. If that were the only consideration, venting would certainly be a minor constructional feature. There are empirical rules for the size of port which offer the mouldmaker a ready guide for adequate venting. But what about design features such as replacement, part withdrawal or material retention, which should be given equal regard. When considering venting, the designer should pose the following questions:

(i) Is the vent tube an expendable item; if damaged or blocked can it be replaced easily?
(ii) Is it possible to ensure the vent bore is clear after each production cycle?
(iii) Is the vent position a hindrance to part removal?
(iv) Is the end of the vent tube likely to be totally obstructed during mould rotation?
(v) Will material be lost through the vent tube bore?

In practice many of the vent tube problems are overcome by use of flexible PTFE tube (Fig. 28a). Nevertheless, rigid steel, purpose-designed vent tubes provide reliable constant trouble-free production and should occupy a prime position within the overall specification of a well designed mould (Fig. 28b).

6. MOUNTING

Earlier the importance of the mould parting flanges and how they formed the key feature to the production of a well constructed mould were considered. One of the most common assumptions made is that only when the mould has been received in the moulding shop can the decision be made as to how and where it is to be mounted onto the machine.

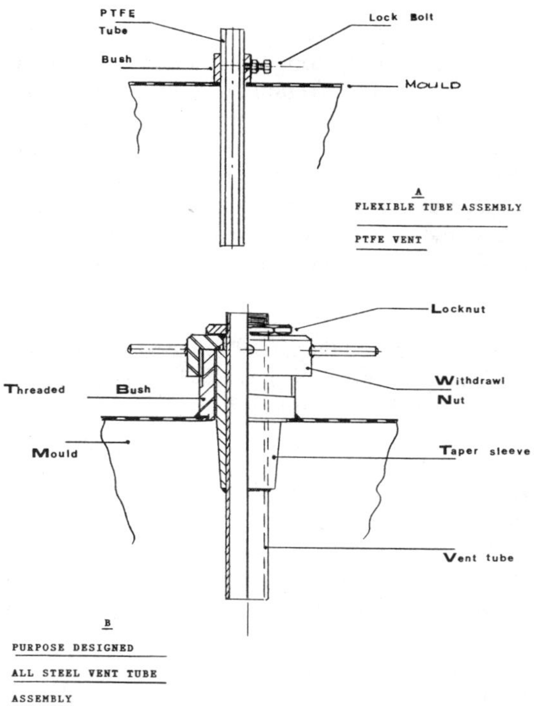

FIGURE 28

So often this approach has led to heated conversations between the moulder and the toolmaker about mismatch and misalignment. It must be realised that:

(a) Indiscriminate welding of mounting frames to the mould flange will produce misalignment.

(b) Supporting the mould entirely at the flange, particularly with deep heavy moulds, is likely to create bending or flexural stresses at the flange joint, as shown in Fig. 29.

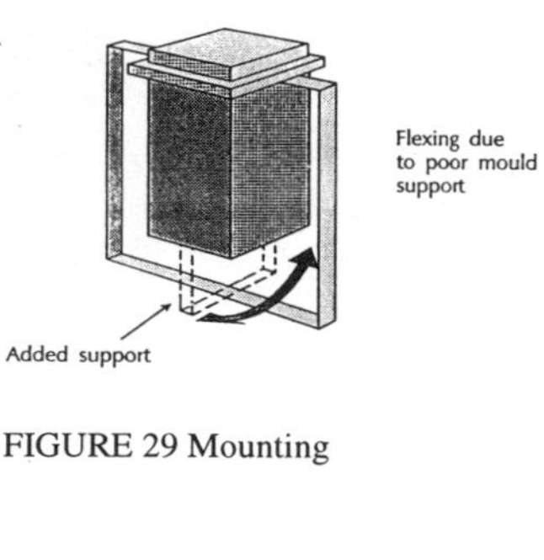

FIGURE 29 Mounting

It is by far the best policy to consider mould mounting as part of the design exercise. Allow attachment points to form part of the tool construction and provide the attachments with drilled holes. Then if the final mounting or grouping is required as a production decision, the framing or support legs can be bolted in place without affecting the mould-half alignment. There is little reward in allowing the toolmaker to carefully construct, over a period of weeks, a perfectly formed, reverse, replica of a product only for it to be deformed by a few minutes of carelessly applied weld. Thus, all additional features such as mounting points, handles, jacking studs, side locations should form part of the initial specification to the mould maker.

7. MAINTENANCE

Once a mould has been commissioned it should be part of all good moulding practices and quality assurance procedures to carry out post-production checks. This can be set up in the form of a standard check list as follows.

Areas for Attention:

1. Check for distortion or damage which is impairing:
   (a) Release (minor problems can escalate)
   (b) Finish.
2. Check all bolted fixtures and tighten where necessary. If seizing is a problem, then brass or stainless steel nuts on steel bolts is a solution (or use a proprietary brand of anti-seize compound).
3. Check all clamps and re-adjust to current loading if necessary.
4. Look for areas of material leakage and 'Build Up'. Clean and re-bed surfaces if necessary.
5. Check all loose inserts for fit and damage.
6. Check vent.
7. Check flange fixing for any undue signs of stress (attachment).
8. Look carefully for metal thinning - erosion or corrosion.

The mould forms an integral part of any production plan and therefore should be maintained in good order if production standards are expected to be kept. This is an accepted practice for Injection and Blow Moulds. Although rotational moulds are constructed in a different manner, with differing tooling skills, and are generally less expensive, these are not reasons for neglecting the good practices of maintenance and regular inspection.

CHAPTER 6
Cast Aluminium Moulds

S. Scaccia

There are several integral parts to the rotational molding process all of which are dependent upon each other to produce a quality product. This chapter will discuss the area which is the most prominent relative to the shape of the plastic part. Most specifically we will discuss how cast aluminum tooling is designed and constructed to provide the foundation for forming the plastic part in the process.

1. HISTORY

Cast aluminum tooling has been in existence since the beginning of the Rotational Molding industry. Early versions were cast in sand using the typical cope and drag technique. This was the only available form of aluminum casting at the time and consequently was used exclusively.

In the 1950's a new aluminum casting technique was introduced to the rotational molding tooling industry which utilized plaster instead of sand. This technique, called plaster casting, offered the following advantages over its completely sand cast counterpart:

1. Thinner walled castings for faster and more direct heat transfer. Sand castings are typically to ⅝ to ¾" thick whereas plaster castings are ¼ to ⅜".
2. Much lower levels of porosity in the castings are achievable due to higher metal quality made possible by the use of plaster.

Both techniques are still in use today. Sand cast tooling is used primarily in areas where cost is more important than the aesthetic quality of the plastic part and where tools are subjected to very forceful handling in the molding process. Plaster cast tooling is used where the quality of the plastic part is of concern and when capturing the maximum amount of detail is essential.

2. PRODUCT DESIGN

The design of the rotational mold begins in the design stages of the plastic part. There are considerations in the product design which will be dependent upon the mold and vice versa.

When designing plastic parts a designer usually begins with a conceptual idea and produces a sketch or drawing. Once he is satisfied with how the product looks and how it will perform its intended function, he must then consider the effect his design will have on mold design and its ability to be molded. A consultation with molders and mold makers is necessary to address these issues. In this consultation session the following items should be considered:

* Does the design contain any severe undercuts in the direction of part removal? If so they must be designed out or addressed with more mold parts (cost will be added to the mold). Some light undercuts can be tolerated because of plastic shrinkage but the molder and the mold builder must determine, for this design, how much can be tolerated without the need for extra mold parts.
* How many pieces will the mold contain? Standard is two, but if the part has a complicated design which cannot be changed more mold pieces will be required. This will have an impact on the cost of the mold, the maintenance of the mold and the individual piece cost of the plastic part since the operators will be required to handle more mold parts.
* What finish is desired for the plastic part? Will this be a cast-in surface on the mold or will it have to be added mechanically alter the mold is cast? Sometimes the specified finish can have an effect on the performance of the mold in hard to remove pieces.
* Is there an area of the plastic part that is going to create plastic flow problems between the two walls? Does it extend for a long distance (6 inches or more)? If it extends for a long distance, finishing of the mold will be difficult and will add to the cost. If it is very thin, it will be difficult if not impossible to get the plastic resin to flow into the cavity and mold properly.
* Does the design include radii generous enough to allow for the free flowing of the resin into all areas of the mold? How much is enough radius?Again, the experience of the molder and the moldmaker must be called upon to make this determination.
* Are there any molded-in inserts for attaching hardware to the plastic part? What are their locations and are they able to be serviced during the molding operation? (more on inserts later).
* Where will the parting line of the mold be placed? Sometimes aesthetics of part design and molding practicality are at odds on this.
* If compression strength through the wall of the part is important, will strengthening ribs or kiss-offs be designed into the part? (Kiss-offs will be discussed in more detail later.)

As you can see, there are several factors of the part design which have an impact on the mold design.

3. THE PATTERN

After the design has been reviewed for tooling considerations a pattern must be constructed. All cast aluminum tools must begin with a pattern which represents the configuration of the plastic part but is larger, to provide allowances for shrinkage of the aluminum during casting and of the plastic part during molding. Typical aluminum shrinkage is 1.1% and the plastic can shrink anywhere from 0% to 3%. To avoid costly errors the patternmaker must receive an accurate and detailed print of the part prior to construction of the pattern.The pattern must be produced in a suitable material which in the majority of cases is a good workable hardwood. Other media such as plastic, resin and fillers are also used. During pattern construction other molding problems may become apparent from observing a three dimensional object that were not seen in the two dimensional sketch or drawing. These must be corrected even though the pattern may be completely constructed since it is much easier to revise the pattern than the mold once it is cast. For this reason, the designer and molder should approve the pattern for both design and moldability prior to manufacturing the mold.

4. MOLD DESIGN

Several items pertaining to mold design have already been mentioned in previous sections of this chapter. These and others will now be discussed in greater detail.

Parting Lines

The parting line of the mold is the point where the mold pieces meet. The location of the parting line is critical because there will be a witness line at the point of contact of the mold pieces. It is desirable to keep this in an area where aesthetics are not as critical as others yet maintain the moldability of the part. When design of the plastic part requires more than two mold pieces this criterion becomes even more important.

In addition to the location, the type of parting line must also be considered and decided upon. There are basically two types of parting lines in use today, flat with steel keys, and cast-in tongue and groove (see figure 1). Each has its own advan-

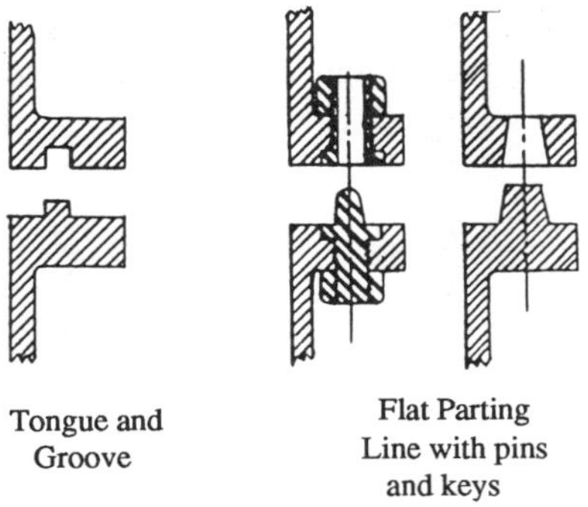

FIGURE 1

tages that will benefit the molder. Consequently he is the one who should make the decision on this. Flat parting lines are level or flat across their entire width. Some molders prefer to use this type of parting line because it is easier to keep clean of spilled resin from the loading procedure. Mold halves are kept from shifting using machined steel guides and pins. Tongue and groove parting lines have a male tongue on one half of the mold and a groove on the other side to accept the tongue and hold the mold halves in place. This type of design is much more resistant to mold shifting during the heating cycle and is much more effective for irregular shaped parting lines that are not in a single plane. Because of its irregular shape across its width the tongue and groove parting line is more difficult to clean between moldings. For this reason the resin is usually poured into the male side of the mold, to prevent spillage on the grooved area of the parting line.

Removable Cores

When an undercut design cannot be removed from a plastic part design because of critical aesthetics or function, removable cores must be designed into the mold. When the undercut is a protrusion into the side of the piece, a removable part must be made for the tool. This part must be attached to the mold prior to the resin filling operation and removed prior to demolding the plastic part on every cycle of the mold. The added labor to do this is the cost of designing in undercuts to a plastic part. Usually this removable part is placed on a seating area of the mold and simply clamped into place. For undercuts that contain threads such as filler necks, the core must be unscrewed from the molded plastic piece prior to demolding. When the undercut is a through-hole the core can be a fixed part of the mold or removable, as desired by the molder.

Ribs or "Kiss-offs"

In order to strengthen the sides of a hollow plastic part ribs or kiss-offs are designed to extend through the hollow wall of the part and connect to the inside of the opposite wall, creating an internal rib in the part (see figure 2). Since the rib or kiss-off touches the opposite inner wall, during the melting and forming process, the kiss-off becomes an integral part of the finished plastic part. Mold design plays a critical part in the ability to effectively utilize this design criterion. The mold

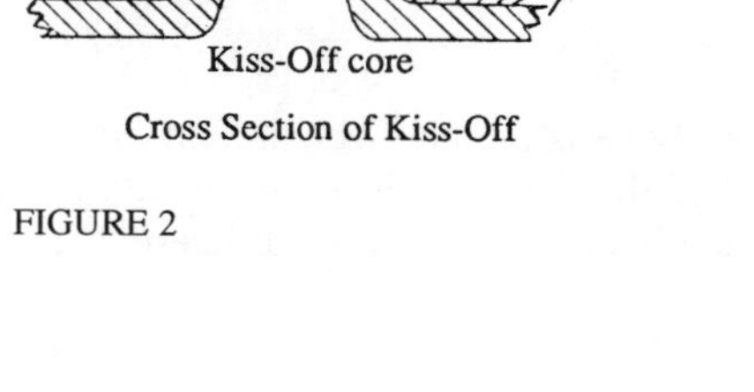

Cross Section of Kiss-Off

FIGURE 2

must provide for enough material to get to the area of the kiss-off and at the same time maintain a distance between the upper mold cavity and the lower mold cavity to provide for a solid wall of plastic to form between them. A variation of the kiss-off is the kiss-through, where the kiss-off part of the mold provides for a hole to be formed in the opposite side of the part by penetrating through the opposite side of the mold. This design is useful in providing strength and allowing for rain or other water drainage from horizontal walls. Either of these kiss-offs must be designed with sufficient draft to allow the plastic part to be removed easily. In addition the kiss-offs must not be located in an area which will be difficult to get heat to, such as in a deep core of the mold. This will cause improper material build up and the two sides will not touch, thus eliminating the strengthening effect of the kiss-off. It should be noted that this procedure cannot be done without some noticeable witness spot on the plastic part. If aesthetics are important perhaps kiss-offs should not be used.

Molded Inserts

Many rotationally molded parts are themselves part of an assembly. As such they are required to be attached to other parts, or other smaller assemblies must be attached to them. In either case it is not always practical to do so by screwing directly into the plastic part. To provide the necessary strength to hold the parts together, metal threads and bolts are used. The threads are provided by a metallic insert that is molded into the plastic part, thus becoming a part of it. These inserts come in various sizes and shapes depending upon the load requirements and the type of fastener being used. The mold must be designed so it is able to hold the inserts in place during the molding cycle and release them after the part is formed. To accomplish this, two methods are employed. The first, which is widely used in the United States, involves drilling a hole through the mold wall in the location of the insert. The hole is then machined and a hardened bushing is inserted into the hole to prevent the aluminum from wearing at this point. This is then ready to receive a bolt which has the same thread as the insert. The bolt is placed into the hole from the back side of the mold and the insert is screwed onto it, thus holding it in place. After the part has formed, the bolts must be unscrewed prior to removing the plastic part from the mold wall. Since the plastic part shrinks upon cooling it will pull away from the mold wall. If there are bolted in inserts present, the part will not be permitted to shrink in the area around the insert so warpage of the wall will occur. In order to prevent this a spring is placed onto the bolt prior to inserting it into the mold and only tightened to compress the spring lightly. When the part shrinks, the insert will pull the bolt against the spring and move with the part. In Europe it is a widely accepted practice to utilize strong magnets to hold the inserts in place. This technique still requires machining of the mold to install the magnets but it eliminates the necessity of using bolts and springs. Since they do not utilize any mechanical attachment of the inserts to the mold, the magnets allow the part to shrink freely away from them, again minimizing warpage. There have been several variations of these two techniques; however, the principle is basically the same.

Surface Finish

Here the designer must decide how he wants the finished part to look aesthetically. The surface finish of the part will dictate how the mold is to be finished or treated. Several options of finish are available to the designer from very smooth to very rough. In the rotational molding process, all plastic part finishes are put in by the mold. In a few cases there is some post molding work done but this is very infrequent. With the exception of non-skid surfaces all finishes are put in for their aesthetic value. Examples of molded in finishes are: shot peen, sand blast, hand sand, leather grain, etc.

A note on PTFE coatings: There has been considerable experience in using a non-stick industrial coating on the surface of the mold. These coatings are primarily used for their release properties for high volume production but they do impart slightly more of a glossy surface to the finished part. This finish comes with some difficulty as the coating can cause the plastic part to shrink more than an uncoated mold, and there is more of a chance for warpage. In the end, coatings should be considered for their release properties first, then the resulting surface properties.

Number of Pieces

In the simplest (and most desirable) form the mold will have two halves, a top and a bottom. In some cases however, the geometry of the finished part will not allow it to be pulled from a simple two piece mold because of severe undercuts which cannot be accommodated with the use of removable cores. When such a case exists, the mold builder must utilize a multiple piece technique to construct the mold. The number of multiple pieces can be as high as six, seven or more; however, it is in the best interest of all involved to keep the number at an absolute minimum. Increasing the number of pieces of the mold has the following disadvantages:

1. Higher price of mold, framing and production. In addition to the extra pieces to cast there is also the additional work of constructing a parting line and all the necessary frame work to make the extra piece fit on the tool.
2. May cause longer lead times for the tool. This is critical when getting a product to market depends upon a definite starting date.
3. Higher risk of premature wear. Especially around the parting lines.
4. Higher part production costs. It takes longer to service a four piece mold than a two piece mold and the molder may not have the option of running the part on a multiple mold frame, therefore increasing the piece part price.
5. Higher maintenance costs due to the higher risk of damaging parting lines or surface finish.

Due to the disadvantages mentioned above many molders will make an effort to convince designers to change the design of parts so that they may be run on a two piece mold.

5. FUNCTION AND OPERATION OF MOLD

Once the design of the mold has been established, the functionality of the mold must be considered. It is here that the designer normally has little input, so the functional aspects of the mold are established by the molder and the mold builder. Each molder uses his own knowledge in establishing the way in which he is going to run a mold and will specify certain mold building criteria in order to fit his personal manufacturing requirements. The following considerations must be made:

1. What size and type of machine will this mold be run on?
2. Will it be run once and evaluated? If so a temporary mount should be provided.
3. Does he prefer to have bolts or clamps to hold the mold halves together?
4. Are there any deep cores that will create difficulty in getting heat to that area of the mold?
5. What type of parting line is most advantageous?
6. What would be the best size vent, and best location for venting?
7. What percent shrinkage must be factored in?
8. Will duplicate molds be required? If so, cast tooling will be advantageous due to the repeatability that is not possible with fabricated tools.
9. Will shielding or varied wall thickness be needed to accommodate thick or thin part wall requirements?

Let's look at some areas that these considerations will affect in the building of the mold.

Framing

Next to the design and condition of the mold the frame is the most important feature of the operation of the mold. The frame will be utilized on each cycle of the machine and must withstand the rigors of repeated heating and cooling cycles and the constant mechanical abuse it will receive in the molding and demolding process. It provides strength to the mold and a means for clamping the mold pieces together. It also provides for the attachment of the mold to the molding machine.

In sand cast molds the frames are attached to the mold via fabricated channels or bars that will bolt to the parting line. This requires that the parting line be countersunk to accept the head of the mounting bolts and still allow the parting lines of the two mold halves to meet. In some cases these framing members are welded to the mold avoiding the machine work in the parting line. Many times in the sand casting mold gussets are cast into those areas which will accept frame members to give the mold some added strength to withstand the abuse of operation.

In plaster cast molds the frames are mounted to rods or posts that are cast into the mold on the back side of the parting line. These posts also provide a service during the casting operation since they act as risers to help ensure good metal quality. Prior to framing, the posts are cut off at a specific length and are drilled and tapped. The frame members will then be bolted into these threads. Usually one half of the mold is mounted with springs between the mold and the frame to accommodate any unevenness in the mount and allow for some give to ensure a snug fit of the parting line around its entire perimeter. Unlike the sand cast molds, the plaster cast molds use square tubing for their frames which provide a greater degree of strength and stability. In addition they allow the attachment of other devices such as clamps, pry pads, or mounting devices for sliding cores or other moving mold parts.

As stated earlier the frame must also provide for a means to attach the mold to the machine. This is done through the attachment of a mounting plate which is welded to the frame and is fabricated to match the mounting design of the intended machine. In many cases this can be provided in a form of a temporary mount or provisional mount to allow the molder and designer to mold a few preliminary parts for approval and then send the mold back for any revisions they feel are necessary after seeing the finished plastic part. A temporary mount consists of an H frame with a mounting plate on the male side, and a lifting hook on the top side. The mold is run with vice grips or suitcase clamps that hold the mold pieces together.

It should be noted that each piece of the mold must be mounted to a frame separately and then fitted together. The selection of what piece of the mold the mounting plate will be welded to is determined by: 1) which piece of the mold will be the easiest to fill with resin without spilling on the parting line or missing it completely and 2) which piece will provide the most resistance to demolding of the part. This then will be dependent upon what method the molder uses in removing a stubborn part from the mold. He may either want that piece to stay attached to the machine for leverage or have it pulled away to be easier to work with.

When molds are large, frame placement becomes an issue. The capacity of a rotational molding machine is determined by two factors: the weight of the molds and their frames, and the size of the sphere the mold and frame creates as it rotates inside the machine. If a mold is approaching the limitations of the intended machine then the frame cannot be mounted so it increases the diameter of the combined frame and mold. This requires the utilization of a technique whereby the cast in posts in the case of plaster cast molds, or the welded on frames in the case of the sand molds, must be shorter than the mold is tall, and the frame, after attachment, does not exceed the longest dimension of the mold.

Spiders

The method described above is used when there is one mold to be run. Basically each time the frames are separated one part is removed and the cavity is filled with resin to start the next cycle. In many instances multiple molds must be run to maintain large production requirements and to keep the labor involved in opening and closing the molds to a minimum. When this is the case the molder turns to a Spider or multiple mold frame. The term spider comes from the shape this frame takes as it stretches over several mold halves with frame members that resemble a spider web. Spiders are designed to accommodate several molds at a time so that when they are opened, all mold halves attached to them are opened at once. This allows a machine crew to service all molds in the spider at once, speeding up the loading and unloading time, thus increasing productivity. Spiders are used whenever high volume production is a must.

Clamping Devices

Clamps come in various sizes, shapes and designs but for rotational molding, clamps usually fall into two categories. They are either of a threaded variety (bolts) or they are non-spring lever type clamps. Clamps are used primarily to hold mold halves together but also get involved in other areas of the mold as well. Removable cores and filler ports are two other areas in which clamps are used.

For sand cast molds, besides the use of clamps to hold mold halves together, bolts are used. These penetrate the parting line and are threaded into a nut on the other mold half usually welded into place to facilitate tightening and prevent loss.

In the case of plaster cast molds the clamps are attached to the frame members so that the mold does not see the abuse of the constant tightening and loosening of the clamps upon each cycle. Clamps are primarily used on individually mounted molds. On spiders, bolts are the primary clamping device.

Heat Transfer

Since the rotational molding process requires heat for the forming process, the rate and evenness of the heat transfer from the ambient air through the mold and eventually to the resin is the most critical part of the process. Plastic part design and consequently mold geometry does not always provide for even heat transfer. Some molds have deep cores which tend to shield a part of the mold from the heated air inside the machine furnace. If there is insufficient heating of that part of the mold it will result in a very thin plastic part in that area and be cause for rejection. Mold design can help alleviate this situation by providing a network of heat absorbing fins that have been cast right into the mold in the deep core areas. These fins provide more surface area for the heated air to cross over, resulting in more heat reaching the mold in that area.

6. MAINTENANCE

Several areas on the molds and frames will require general maintenance from time to time.

Parting Lines

This is one area of the mold that is the most important and is subjected to the most abuse during normal operations. With each mold cycle the mold halves are subjected to impact against their mating halves. The machine operators responsible for opening and closing these molds may not always take the necessary precaution to ensure the mold halves are properly aligned prior to assembling them. The result is damage to the parting line at the point of contact. Although each incident is minor, repeated activity will cause permanent damage. When severe damage has occurred, the mold should be returned to the mold maker for repairs. The refurbishing of parting lines requires skill and experience and is best left to the experts.

Another source of damage to parting lines occurs when resin is accidentally spilled onto the parting lines during the filling of the mold. If it is not removed prior to the mold being closed and clamped for the next cycle, the result will be a melted-on resin layer on the parting line. If this happens, mechanical scraping of the parting line is required to remove the resin from the parting line. If this is not done the mold halves will not fit together properly, resulting in flash at the parting line of the parts. The tools used, the technique, and the frequency of this scraping will definitely affect the integrity of the parting line.

One final area of concern for the parting line is in getting the mold halves apart once the plastic part is formed and it is time for demolding. On occasion plastic parts will stick in the mold halves causing difficulty in getting the mold halves apart. Although steel pry points are provided for the purpose of assisting the machine operator in getting the mold halves apart, they occasionally will pry directly on the parting line with a steel bar when attempting to open a mold. Abuse of this type is catastrophic on the parting lines.

In summary, most problems associated with parting line wear and maintenance can be avoided by proper training of operators and care during the molding operation.

Clamps

The mold clamps are mechanical devices that are also subjected to abuse during the molding operation. These clamps must be engaged and disengaged every cycle of the mold. Hammering or leveraging clamps to get them open or closed can be avoided by keeping the clamps in proper adjustment. Should mold clamps become defective, it is a simple replacement operation that can be done by the molder.

Cleaning

There are no specific rules to govern the proper cleaning of rotational molds. It would be best if the operators and maintenance personnel would bear in mind that these molds are made of aluminum, which is a relatively soft metal, when selecting the proper tools to be used in the cleaning operation. The maintenance person should avoid sharp hard tools and harsh chemicals that would damage aluminum. Molds should be cleaned thoroughly after use and prior to storage.

CHAPTER 7
Design of Rotationally Moulded Products

G.L. Beall

1. DESIGNING ROTATIONALLY MOLDED PLASTIC PRODUCTS

The design of all successful products begins with the designer's determination of the functional requirements of the product in its end use environment. If those requirements include a need for the early market introduction of large, hollow plastic parts of irregular shape, in small to modest quantities, at a low initial financial investment, the designer should think of rotational molding.

The rotational molding process is a designer's dream come true. This process provides the designer with more design freedom than any other one plastics manufacturing technique. Rotational molding can provide both closed or open hollow parts. Single wall and closely spaced double walls can be produced. The walls can be thick or thin, solid or foamed, and fiber can be added for reinforcement. The parts can be transparent, or a full range of colors can be provided. The shape of rotationally molded parts can be as simple as a play ball, or as complex as a farm tractor instrument panel. Rotational molding is a "designer friendly" process.

Once the design engineer has a clear understanding of the requirements of the product, the next step is to start hunting for a shape that will satisfy those requirements. This search normally takes the form of simple sketches. What the designer is actually doing is experimenting with different shapes in sketch form, trying to find the optimal structure that will satisfy the product's functional requirements. The poolside floating chair, shown in Fig. 1, is a typical concept sketch of this type.

In this illustration, the upper lefthand sketch is made up of two simple molded six-sided boxes for floatation. The armrests and legs are provided as added metal hardware. This design requires two cavities and two molding operations, plus assembly.

The upper righthand sketch shows a combined seat and back, plus four molded legs. This design requires two cavities and five molded parts, plus assembly. The removable legs would significantly reduce shipping volume and storage space.

The lower lefthand sketch shows a one-piece molded chair. This concept minimizes the possibility of weakness between the chair seat and the legs in the two

Some drawings and data reprinted with the permission of the Association of Rotational Molders (ARM).

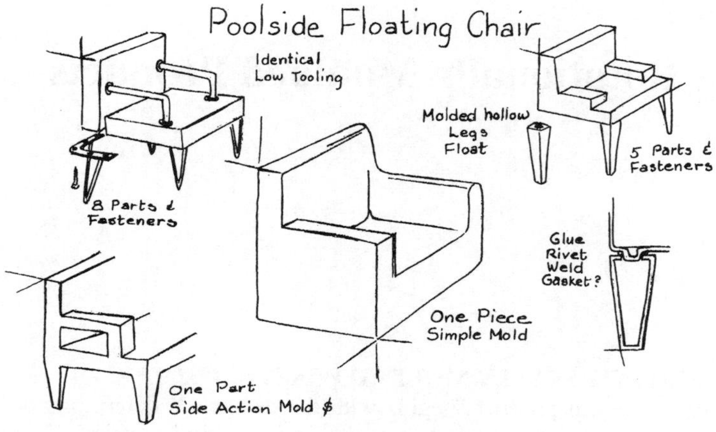

FIGURE 1 Typical new product concept sketch

assembled concepts. This design can be produced with one cavity and one molding operation. The cutouts under the armrests require a more complicated mold with loose parts.

The center view shows a one-piece chair. This part can be produced in a simple, low cost, two-piece mold in one molding operation. All of the hardware and loose pieces required with the other concepts have been eliminated. This concept has no metal parts that could be attacked by the chlorine in the swimming pool water. There is no longer any concern for the strength of the legs. All things considered, this is the best concept for a floating chair. Unfortunately, the chosen design does not present a pleasing appearance.

The appearance of a large storage tank or a tote bin may not be very important. Any product that people work with or come in contact with requires attention to appearance design and human engineering. Appearance design has to do with the shape, proportion, color and surface finish that allow the product to be pleasing in appearance. The product must, simultaneously, give the impression that it can do what it is intended to do.

The size of the product must be such that it will fit human beings or that can be used easily by people. This kind of design falls under the category of human engineering.

As an example, the two chairs shown in Fig. 2 will perform the same function. The chair on the left, with its simple shape, straight lines and large draft angle, would be easy to design, draw and dimension. A mold could be built at a low cost. Molding of the chairs would be easy. However, this chair looks harsh and uncomfortable.

The chair on the right is not beautiful, but it is more pleasing in appearance. This chair also incorporates some important human engineering factors. The front-to-

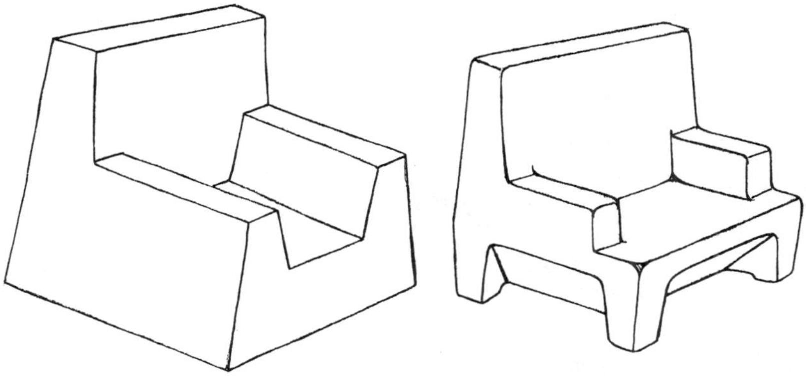

FIGURE 2 Industrial design considerations

back length of the seat has been shortened. The armrests have been lowered. Space has been provided for the user's heels. The combination of all of these little details makes a better looking chair that is more comfortable to use. At the same time, these changes did not have a detrimental effect on the cost of the tooling or the molding of the parts.

The next logical step in the design sequence would be to choose a plastics processing technique that could produce a hollow, one-piece plastic chair of this shape. This floating chair was designed with rotational molding in mind. However, there are other processes such as twin sheet thermoforming or extrusion blow molding that should be considered before a final decision is made.

Once the process has been chosen, the designer must then choose a plastic material that will, simultaneously, satisfy both the end use functional requirements and the manufacturing requirements. In the case of rotational molding, every effort must be made to select a material from the four families of plastics that are good for the rotational molding process. These four families of materials are: polyethylene, polyvinyl chloride, nylon, and polycarbonate.

If no suitable material can be found within these four families of plastics, the designer can then consider some of the other plastic materials which can also be molded. In making this decision, the designer must remember that these other materials are not ideal for the rotational molding process. These other materials will be more difficult, and therefore more costly to mold into acceptable parts.

Once the shape of the part has been established and the combination of plastic material and processing technique has been finalized, the designer can then turn his attention to the performance characteristics of that combination of shape, material and process. In other words, will this material, molded in this shape by this process, perform as required. For example, will the chair be strong enough to support a 110 kilo man with a 50 kilo lady in his lap, plus one 20 kilo child and one 30 kilo child sitting on the chair arms? This chair will require strong legs (Fig. 3).

FIGURE 3 Structural analysis of preliminary design

Chairs are bulky and hard to handle. They will require enough impact strength to withstand shipping and rough handling.

It can be anticipated that a poolside floating chair will be subjected to direct sunlight. The intensity of the sun and the resulting ultraviolet light degradation of the material's physical properties and the fading of its color will be different in the United Kingdom than they are in Saudi Arabia. Where will this chair be used?

If the chair is to be used indoors in a public place, there may be additional requirements regarding the flammability of the material.

A complete review of the evaluation of the strength and environmental factors for the floating poolside chair is beyond the scope of this chapter. However, it is worth noting that the evaluation of a rotationally molded part is done in the same way as for any other plastic product. The one important difference is that the physical properties used for strength calculation will be more reliable if they are taken from rotationally molded test plaques, and not from the more common compression molded test specimens.

The next step in the design process is the finalizing of the design and the preparation of the detailed engineering drawing, as shown in Fig. 4.

The part drawing, or blueprint, is an extremely important part of the new product development scenario. The piece part drawing will establish the shape of the part and the dimensional interrelationships that will control how that part fits into the rest of the assembly. The drawing will also contain all of the important little details, such as wall thickness, corner radius, draft angles, reinforcing ribs and surface finish.

It is this same drawing that the tool maker will use to build the cavity which will actually produce the molded chair. The cavity can only be as good as the part drawing.

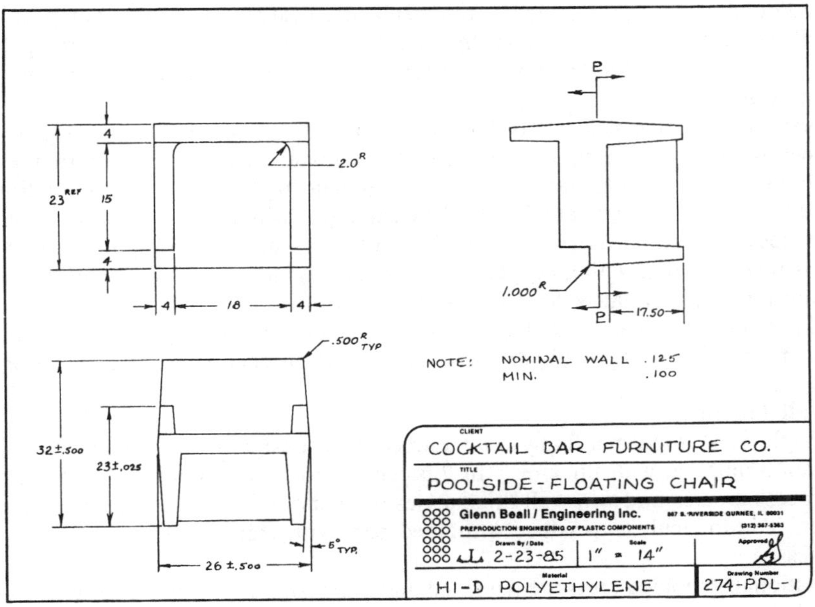

FIGURE 4 Chair part drawing

The molded parts can only be as good as the cavity that formed them. The ultimate success of the new plastic product is, therefore, controlled by the quality of the part drawing. It is in this area of the part design that plastic products begin to fail.

The product design community was quick to recognize the design freedom provided by an easily formable material such as plastic. Plastic product designers have made tremendous strides since the Second World War. The size, complexity and quality of many contemporary plastic parts would have been unthinkable ten years ago.

Each combination of plastic material and process has its own unique capabilities and limitations. Product designers welcomed the design freedom of plastic materials and processes. However, they were less enthusiastic about acknowledging the limitations and design requirements of plastic materials and processing techniques. As a result, the plastics industry is simultaneously blessed and cursed with excellent product designs which are defined in part drawings that are less than optimal. The cost and quality of the resulting products suffer accordingly.

Considering the fact that plastic piece part design is not being done as well as plastic product design, the rest of this chapter will concentrate on the design guidelines and rules of thumb for rotationally molded plastic parts.

2. ROTATIONALLY MOLDED PLASTIC PART DESIGN

A good quality piece part design is the result of a designer's understanding of the capabilities, limitations and unique requirements of the material, the tooling and processing technique that have been chosen, coupled with a meticulous attention to design details.

Rotational molding is a relatively new plastics processing technique. A product designer who is contemplating the use of rotational molding for the first time should proceed with caution. One of the reoccurring problems with rotational molding is that many parts are designed by engineers with no prior experience with the process. First time users of rotational molding will enhance their chances of success by consulting with an experienced design engineer, molder or material supplier during the preliminary design phase of a new project.

A little additional care and attention in the part design phase of the new product development cycle will pay big dividends when that product goes into production.

2.1 Wall Thickness

The single most important thing to remember about designing any plastic part is to maintain a uniform wall thickness. In this regard, rotational molding is unique, as the process creates uniform wall thicknesses no matter what the designer specifies. In fact, it is difficult to produce a molded part with significant changes in wall thickness.

Rotationally molded parts are designed and dimensioned on only the outside surfaces that are in contact with the inside surface of the cavity. The inside of the molded part will be freeformed and its size and shape will be dependent upon the outside size and shape, minus the wall thickness.

The type of plastic material used and the thickness of the nominal wall will determine the strength and loadbearing capability of the finished part. The wall thickness required to sustain a given load can be determined by the conventional techniques that are applied to other plastic parts However, as molded, physical properties should be used in these calculations.

One of the special characteristics of the rotational molding process is that there are no forces which push the plastic material through the cavity, as is the case with compression and injection or blow molding. In rotational molding, the mold rotates through a puddle of liquid or powdered plastic, and the plastic adheres to the hot inside surfaces of the cavity to build up the desired wall thickness. Considering this situation, the ideal design for a rotationally molded part is a hollow shape, where the various elements in the part design are smoothly blended from one contour to the next.

The lack of force and the absence of shear in the plastic material make the design of rotationally molded parts very dependent upon the particular plastic material being used. As a result, the parts must be designed to accommodate the flow properties of the material.

The relatively long, high temperature heating portion of the rotational molding process can be tolerated better by some plastics than others. Thicker walled parts require longer heating cycles. The possibility of thermal degradation limits the maximum wall thickness that can be produced with a given plastic material.

There are always exceptions, but the following maximum and minimum recommended wall thicknesses can be used as a guide to what wall thicknesses are practical.

Polyethylene provides the designer with the widest range of wall thicknesses. Polyethylene parts will be at their best with a wall thickness in the range of 1.5 to 13.0mm. Wall thicknesses of as thin as 0.5mm have been successfully produced; 50.0mm thick walls have been molded on large storage tanks. A wall thickness of 25.00mm is not uncommon. However, these thick walls result in long molding cycles.

Polyvinyl chloride (plastisols) have been reliably molded into small medical devices with wall thicknesses of less than 0.25mm. Most polyvinyl chloride parts are designed with wall thicknesses in the range of 1.5 to 10.0mm.

Nylon parts are generally designed with wall thicknesses in the range of 2.5 to 20.0mm. In special cases, parts as thin as 1.5mm and as thick as 38mm have been molded with satisfactory results.

Polycarbonate is typically used for molded parts with wall thicknesses in the range of 1.5 to 10.0mm.

Rotationally molded parts are hollow. They are produced in molds without internal cores. In many cases, this provides the designer with the important option of changing the thickness of a molded part after the production mold has been sampled and actual molded parts have been evaluated. In most other processes, changing the wall thickness would require extensive mold modifications.

Wall thickness has a direct effect on the cost of the molded part. In addition to the added cost of the material used in a thicker wall, the cycle time and the energy required to heat and cool the plastic will be directly related to the wall thickness. For example, a 0.76mm increase in the wall thickness of a Nylon 6 part will result in an increased heating time of approximately two minutes. One half minute of additional heating will be required for each 0.6mm increase in the wall thickness of a polyethylene part.

2.2 Wall Thickness Uniformity

Rotationally molded parts are produced in molds with no internal cores. As a result, the wall thickness cannot be controlled exactly. Depending on the shape of the part and the material being molded, it is generally agreed that the wall thickness can be controlled to ±20%. If wall thickness uniformity were more important than cost, the variation can be reduced to ±10% in some cases.

In the case of closed molding processes, such as injection or compression molding, the standard way of indicating the wall thickness of a molded part is to specify a specific thickness and an allowable tolerance; for example, "3.25±0.2". Unfortunately, this same drafting practice is also applied to rotationally molded parts. A better method of specifying the wall thickness on a rotationally molded part is to indicate the nominal wall thickness and the minimum allowable thickness that will be accepted anywhere on the molded part. For example, "nominal wall thickness 3.25mm; minimum allowable wall thickness 2.60mm".

In many cases, the wall thickness is monitored and controlled on a production basis by weighing the molded part. Obviously, this approach can only be used after the production mold has been built and the weight of acceptable molded parts has been established.

The inability to specify a specific wall thickness, or to control the inside dimensions on a rotationally molded part, is unsettling to a designer who is accustomed to working with only closed molding processes such as injection molded structural foam or reaction injection molding. However, extrusion blow molding and twin sheet thermoforming share the same limitations. In fact, rotational molding maintains more uniform wall thicknesses than either one of these other competitive processes.

2.3 Closely Spaced Parallel Walls

One of rotational molding's advantages over other processes is the ability to easily produce double walled hollow parts, such as one-piece insulated ice chests or the boat shown in the cross section illustration, Fig. 5. Many of these hollow parts with closely spaced parallel walls are filled with self-rising foam to provide additional stiffness or insulation.

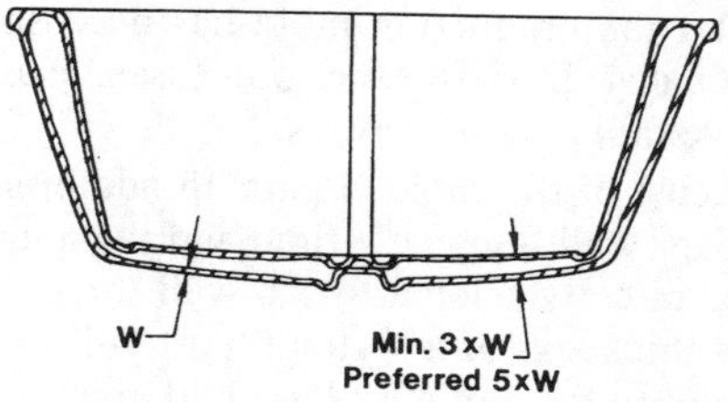

FIGURE 5 Design rules for closely spaced walls

The bulk factor of finely ground rotational molding powders is approximately three times greater than the solidified molded part. Considering this fact, the designer must provide a minimum average distance between parallel walls of three times the nominal wall thickness. A smaller open space between the walls of the cavity would not provide enough volume for the powdered plastic. Walls this close together require extra care in putting the powder in the cavity. In some cases, the mold must be closed and the plastic material is then carefully poured into the cavity through a filling port. In extreme cases, the mold must be vibrated in order to get all of the material into the cavity. All of these extra operations add to the cost of molding the part.

There is also a tendency for the powder to bridge across closely spaced walls. This creates a solid wall, or dam, that prevents the powder from freely flowing into other areas in the cavity. This condition can produce nonuniform wall thicknesses in unexpected locations. The bridged over areas also create localized thick sections. Thick solid sections can cause internal voids and external sink marks. These thick sections stay hot longer and shrink more than the surrounding thinner sections. This condition can result in molded in stress and an increase in part warpage.

Wherever possible, the distance between closely spaced parallel walls should be specified at at least five times the nominal wall thickness.

Some rotationally molded parts have been produced with solid, single walls protruding from a hollow part (Fig. 6). These designs have met with mixed results. The liquid plastisols are the best of the rotationally moldable materials for producing nonhollow sections. Wide projections that extend only one or two times the wall thickness can normally be molded. Longer projections bridge over and do not fill out properly. These solid walls will frequently contain large voids due to entrapped air. There are many other plastics processing techniques that are capable of producing single, solid walled parts. However, the rotational molding process will be at its best if all of the details on the molded part are designed as hollow sections.

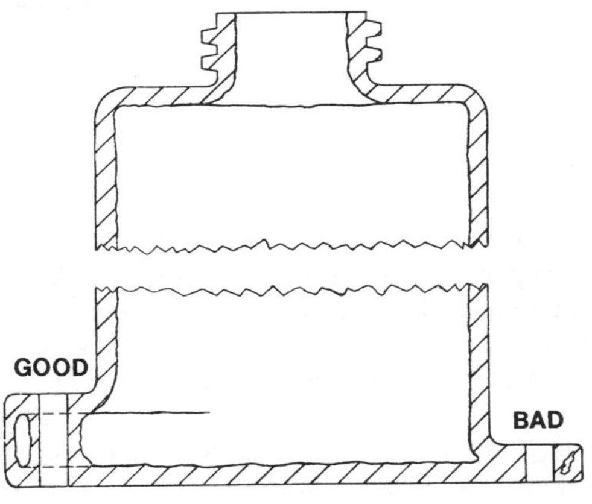

FIGURE 6 Solid walls are undesirable

2.4 Warpage

Rotationally molded hollow parts are produced in a mold with no internal cores. During the cooling portion of the molding cycle, the plastic material can shrink and pull away from the inside surface of the cavity. This condition allows the molded part to warp. The high mold shrinkage crystalline materials are more prone to warpage than materials with lower mold shrinkage factors. There are many tricks of the trade that competent molders use to compensate for part warpage. However, it is difficult and sometimes impossible to completely eliminate warpage in large flat surfaces.

Experienced designers of rotationally molded parts avoid large flat surfaces wherever possible. Where large flat surfaces cannot be eliminated, these sections can be reinforced against warpage by the use of ribbing or crowning. The adverse appearance of warpage can be minimized by adding engraving or texture to disguise the condition. A highly polished or painted surface will accentuate warpage.

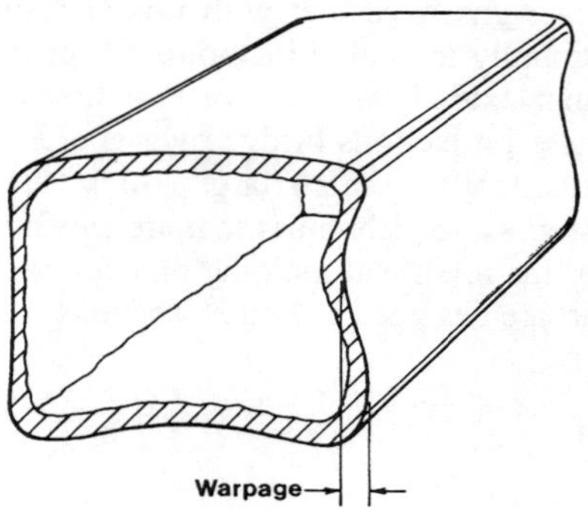

Material Flatness	Ideal	Commercial	Precision
Polyethylene Polyvinylchloride	±.050	±.020	±.010 in/in
Nylon Polycarbonate	±.010	±.005	±.003 in/in

FIGURE 7 Mold part flatness limits in ± cm/cm

The degree of flatness that can be expected on a given part is dependent upon many factors: the size and shape of the part; the plastic material being molded; the quality of the mold; and the way the part is molded. Each plastic part must be considered individually. Fig. 7 lists the ideal, commercial and precision flatness tolerances that can be used as guidelines in designing rotationally molded parts in the commonly molded materials.

The ideal tolerance requires nothing more than normal good quality molding. The precision flatness tolerances can only be achieved at a significant increase in costs.

2.5 Stiffening Ribs

One of the advantages of the rotational molding process is that it has the capability of producing plastic parts with very thin walls relative to their size. These thin walls are made possible by the fact that the plastic material does not have to flow through the length of the cavity. During the rotational molding process, the extremities of the cavity are brought to the plastic material, which remains in the gravity down bottom portion of the cavity at any point in time. This thin walling capability reduces both material costs and cycle time.

Thin walls are advantageous, but many molded parts require greater stiffness than can be provided by the minimum possible wall thickness. In many cases, it is possible to make a desirable compromise between wall thickness, stiffness and cost by incorporating stiffening geometry into the part design. For example, the two six-sided boxes shown in Fig. 8 provide the same function. However, the ribbed shape can provide the same strength as the unribbed part with less plastic material and a shorter molding cycle.

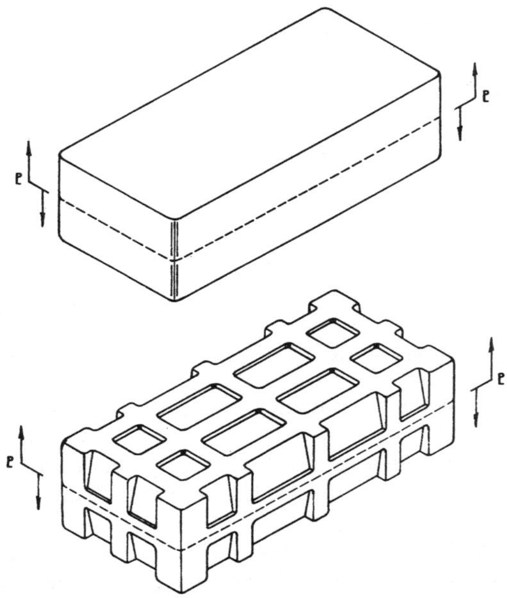

FIGURE 8 Ribbing adds strength while reducing material and molding cost
Reproduced courtesy of the Association of Rotational Molders

Depending upon how the part will be loaded, some consideration should be given to the possibility of hollow ribs of this type failing, due to deforming like an accordian.

Hollow stiffening ribs, of the type preferred for the rotational molding process, are actually closely spaced parallel walls. The minimum distance across a stiffening rib should be at least five times the thickness of the part, as shown in Fig. 9.

The stiffness provided by the rib is determined by the distance the rib projects off of the wall of the part. A rib height of four times the nominal wall thickness will provide a significant increase in stiffness. If this dimension becomes more than four times the part's wall thickness, the width of the rib should be increased proportionally. A failure to provide an adequate width in the rib cavity will prevent uniform flow of the plastic material into that part of the cavity, resulting in bridging, internal voids and an increase in mold shrinkage.

Generally speaking, it is preferable to limit the distance that a rib projects above the wall of the part. A short rib of the type shown in Fig. 9 may not provide the

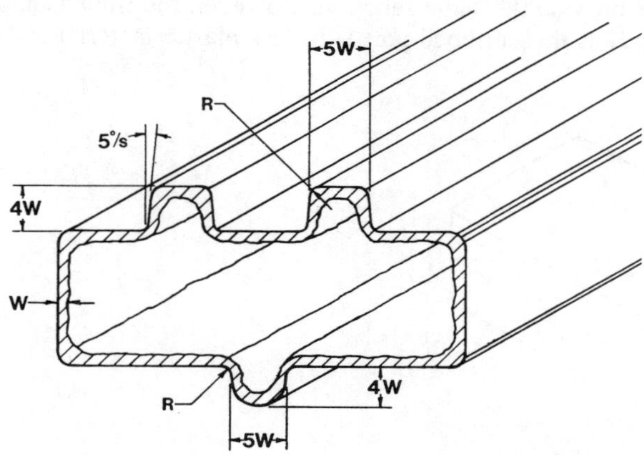

FIGURE 9 Recommended minimum hollow rib proportions

required stiffness. In those cases, it is better to provide a multiplicity of short ribs instead of one tall rib.

In those cases where hollow ribs cannot be used, short solid ribs with the proportions shown in Fig. 10 can be used. These solid ribs are less desirable than hollow ribs, but they can be used as a last resort.

The problem with solid ribs is that the material bridges over and does not properly fill out the rib cavity. If a solid rib cavity fills out completely, it will create an increase in wall thickness in that location. This thicker wall will take longer to cool and it will shrink more than the thinner walls around the rib. The thicker rib is stronger than the thinner walls. When the thicker, stronger rib shrinks more, it can overpower the surrounding walls and contribute to warpage.

2.6 Kiss-off Ribs

Another useful method of providing additional stiffness to a rotationally molded part with closely spaced walls is kiss-off or tack off ribbing, of the type shown in Fig. 11. Connecting the two walls together at the kiss-off creates a very strong box beam type structure.

The exact thickness of the material in the kiss-off area is, in the final analysis, established by trial and error. A thickness of 1.75 times the part's wall thickness has been found to be a good starting point.

While designing parts with kiss-off ribbing, care must be exercised to provide adequate space between two kiss-off ribs to allow the plastic material to freely come in contact with all surfaces of the cavity.

Kiss-off ribbing can be used to produce long stiffening ribs or small isolated areas of added strength. Inner columns, of the type shown in Fig. 12, are often used

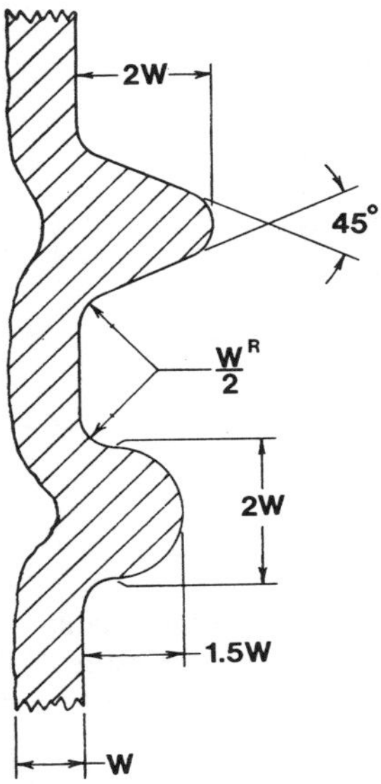

FIGURE 10 Recommended minimum solid rib proportions

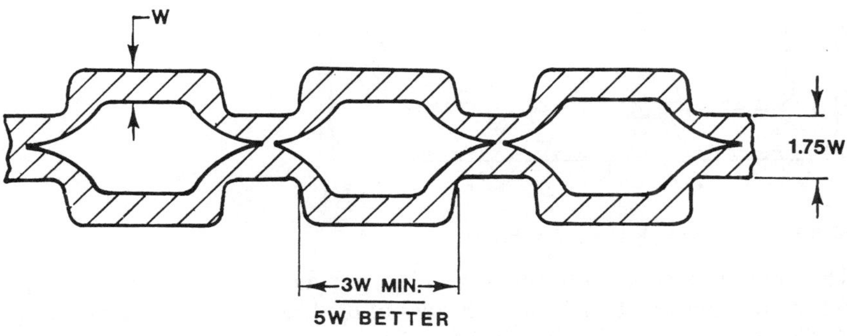

FIGURE 11 Recommended minimum kiss-off proportions

to provide strength, to counteract warpage, or to prevent the sloshing of liquids in portable tanks.

Widely spaced apart parallel walls require kiss-off ribs of a large enough size to allow adequate heating of the cavity in the kiss-off area. As a rule of thumb, inner columns of this type begin to cause molding problems whenever their depth exceeds their width. The proportions of the inner columns shown in Fig. 12 would require special attention in order to ensure adequate heating of the kiss-off area.

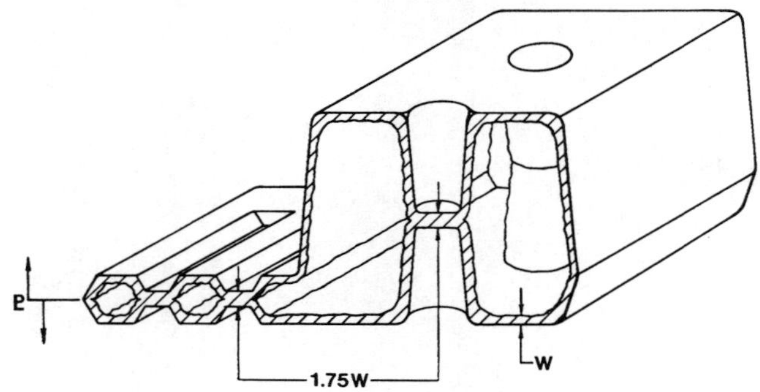

FIGURE 12 Kiss-off rib shapes

Kiss-off ribbing is sometimes used in the bottom of double walled tanks and containers to provide the strength required to support the downward pressure of liquid or granular products. A useful variation of the standard kiss-off rib is the "almost" kiss-off rib shown in Fig. 13B.

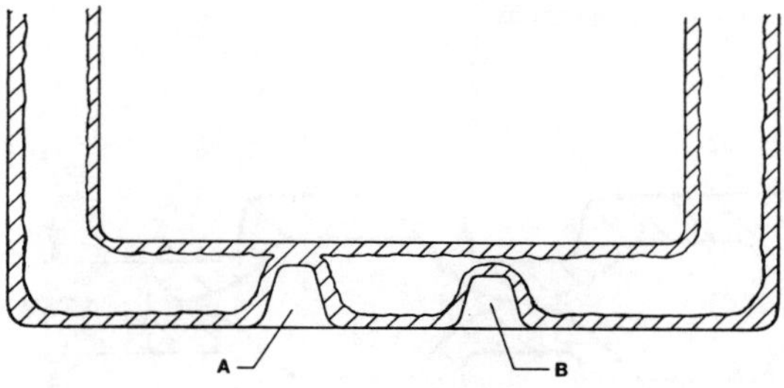

FIGURE 13 (A and B) Kiss-off rib variation

The conventional kiss-off rib, Fig. 13A, provides a lot of vertical support. However, the horizontal walls on both sides of the kiss-off bend in response to the load. This creates a stress concentration at the junction of the supported and unsupported horizontal walls. The "almost" kiss-off rib provides the same vertical support, while eliminating the stress concentration by allowing the horizontal wall to move relative to the vertical support.

A similar kiss-off and "almost" kiss-off are shown being used effectively on the boat, Fig. 5. The conventional kiss-off rib at the keel provides additional strength in the heavily loaded area. The "almost" kiss-off ribs at the sides of the deck allow the inner hull to gain support from the outer hull, while allowing the two hulls to move relative to each other in response to inside or outside forces.

2.7 Draft Angles

All molded plastic parts have to be removed from the cavity that formed them. Plastic parts molded by any process will be easier to remove from their cavities if the part is designed with draft angles on those surfaces which are perpendicular to the parting line of the mold. Anything that makes producing the part easier reduces the molding cost. Draft angles are one of these design features that reduce the cost of rotationally molded parts.

The limiting factor on a rotationally molded part's cooling cycle time is that the part must be cooled enough to have regained sufficient strength to retain its shape after being removed from the cavity. The part must also have regained sufficient strength to resist the forces required to remove the part from the cavity. The liberal use of draft angle, wherever possible, will reduce the forces applied to the part during the demolding process. In reducing these forces, the cooling time, cost, induced stress and part warpage will be minimized.

One of rotational molding's advantages over other processes is that many types of part can be molded straight up and down, with no draft angle at all. This is made possible by the fact that these hollow parts are molded without internal cores. As the part is cooled, it shrinks and draws away from the cavity, which makes it easy to remove from the mold.

Each of the various plastic materials has its own mold shrinkage characteristics. Those materials which have a high mold shrinkage factor, such as polyethylene or nylon, will pull away from the outside cavity wall more than the materials with a lower mold shrinkage factor, such as polycarbonate.

These same high shrinkage factors have the reverse effect on inside surfaces of the cavity, such as the truncated cone and rib detail shown in Fig. 14. Shrinking causes the material to grip these surfaces very tightly.

The softer, self-lubricating materials, such as polyethylene, are always easiest to remove from cavities that have a minimal draft angle.

The strong, rigid materials, such as nylon and polycarbonate, are more difficult to remove from cavities with minimal draft angles.

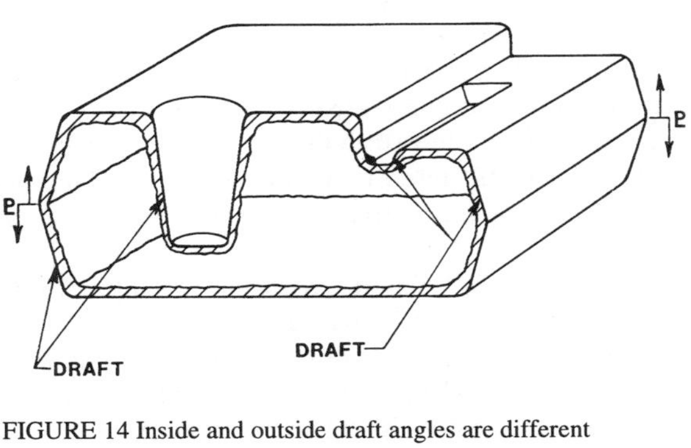

FIGURE 14 Inside and outside draft angles are different

It is difficult to state hard and fast rules for draft angles on rotationally molded parts, as some parts have been produced successfully with no draft angles at all. The ability to produce straight up and down parts with no draft is one reason why designers specify rotational molding.

It is safe to say, however, that liberal draft angles should be incorporated into the design of plastic parts that will be rotationally molded. This is especially true for inside surfaces that will shrink down tightly onto the cavity as the part cools, making demolding difficult.

The draft angles shown in Table I should be applied to rotationally molded parts where their presence will not interfere with the functional requirements of the part in its final end use environment. Parts of large size or unusual shape may require larger draft angles. This is especially true for deep draw parts molded in rigid materials.

Table 1. RECOMMENDED DRAFT ANGLES IN DEGREES PER SIDE

	Outside Surfaces		Inside Surfaces	
Material	Min.	Better	Min.	Better
Polyethylene	0°	1°	1°	2°
PVC	0°	1.5°	1°	3°
Nylon	1°	1.5°	1.5°	3°
Polycarbonate	1.5°	2°	2°	4°

2.8 Surface Finish

The smoothness of the finish on the inside surface of the cavity goes hand in hand with draft angles. As the roughness of the cavity increases, more force is required to remove the molded part from the mold. The stiff materials, such as nylon and polycarbonate, require smoothly polished surfaces on all inside cavity details. This is especially true when only the minimum draft angles have been specified.

All types of molds used for the rotational molding process can be plated, polished, etched, engraved or blasted with shot or abrasive materials.

Many rotationally molded parts destined for the industrial market do not require a high surface finish. These cavities will normally have what is referred to as an "industrial finish". An industrial finish normally means that the cavity will be rough hand sanded. Grinder and cutter marks left from the finishing process may be present. Various mold makers have different interpretations of what an "industrial finish" is. What is meant by an "industrial finish" must be clarified before an order is placed for a new mold.

Photochemical etching is possible on all types of rotational molding cavities. But etching can open up subsurface porosity. In many cases, this possibly can be avoided by the electroforming and aluminum castings' ability to duplicate the required texture off of the original model. Fabricated sheet metal molds can be etched, but the welded areas are difficult to etch with a uniform pattern.

Blasting the cavity with steel shot or abrasive materials of varying size works well with all types of mold. There are some limitations on the blasting process's ability to provide a uniform finish at the bottom of deep narrow recesses.

Textured and blasted finishes represent a multiplicity of tiny undercuts in the wall of the cavity. When these finishes are used, the draft angles recommended in Table I should be increased by one degree per side for each 0.025mm of texture depth.

Fabricated or machined steel cavities will take the highest polish. Cast and electroformed cavities can be polished to a mirror finish; however, the occasional presence of porosity is a limitation.

Cavity finishing can be labor intense. Depending on the size and shape of the cavity and the process by which it was originally made, a highly polished finish will normally be the most costly to provide and maintain.

High quality finishes can add significantly to the initial cost of a mold, as well as the ongoing care required to maintain that finish. Designers should specify only the finish required by that product in the hands of the end user. Care must be exercised to not overspecify the surface finish.

2.9 Undercuts

An undercut on a rotationally molded plastic part is any inwardly or outwardly projecting wall on the molded part that is parallel to the parting line of the mold, where the part must be deformed in order to be removed from the cavity.

An undercut on a rotationally molded part is the exact reverse of a draft angle. Draft angles improve demolding. Undercuts make it difficult to remove a molded part from the cavity. However, undercuts can provide a lot of beneficial functional features on a molded part.

It is difficult to be specific about the design of undercuts on plastic parts which will be rotationally molded. This process, which molds hollow parts with no inside cores, does allow the walls of the molded part to be deformed inwardly to free the undercut. This is one of rotational molding's advantages over closed molding processes, such as injection or compression molding. However, there are limitations on both the design of the parts and the plastic material that is being molded.

The soft polyvinyl chloride plastisols will allow extremely deep undercuts to be forced out of a cavity. The rigid nylon and polycarbonate materials will tolerate little or no undercuts. Polyethylenes are somewhere in between.

The shape of an undercut must be designed to develop a bending force in the direction that the molded part must deform to allow the undercut to be pulled out of the cavity. Right angled undercuts projecting from the wall of a molded part do not encourage that part to deform in this manner.

Referring to the molded part shown in Fig. 15, the potential Undercut A can be avoided by positioning the mold's parting line at Location PL 1. However, this would require that Undercuts B, C, D and E be deformed to force them out of the cavity. All of these undercuts, except the inside Undercut E, could be eliminated by repositioning the mold's parting line to Location PL 2.

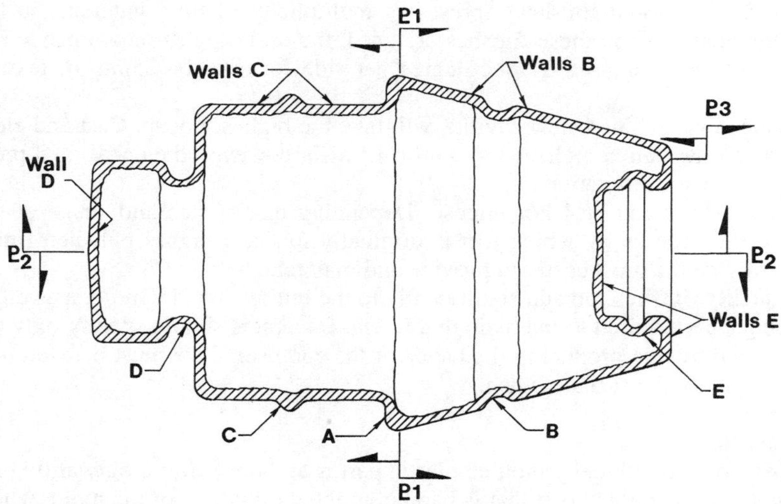

FIGURE 15 Rotationally molded undercuts

It is interesting to note that the soft polyvinyl chloride plastisols would allow all of these undercuts to be easily forced out of the cavity. At the same time, the adverse appearance of the parting line could be minimized by molding this part in a one-piece electroformed cavity with the parting line located at PL 3.

The Undercuts B and C can be stripped from the cavity if the plastic material will deform sufficiently to accommodate the depth of the undercut. Removal from the cavity will be enhanced if the shape of the undercut encourages the inward deformation of Walls B and C.

The Undercut D will be difficult or impossible to remove from the cavity in all but the very flexible plastics, such as low density polyethylene or polyvinyl chloride. The location of Wall D reinforces this undercut and makes it difficult to deform inwardly. The Walls B and C in the previous example are free to bend.

The Undercut E is similar to D. Whether or not it can be stripped from the mold is dependent upon the flexibility and shrinkage factor of the plastic material, the depth and shape of the undercut, and its location in relationship to the two reinforcing walls, E.

Internal undercuts, of the type shown at E, are always more difficult to remove than external undercuts of the same proportions. This is due in part to the fact that, as the plastic cools, it shrinks and draws down tightly onto the cores that form the undercut. External undercuts tend to free themselves from the cavity due to shrinkage of the plastic. Internal undercuts of this type should be avoided where possible.

When they must be specified, their depth must be kept to a minimum and their use limited to flexible or semirigid plastic materials only.

The best rotational molds are of a two-piece construction. However, loose cavity parts can be built into a rotational molding mold in order to accommodate special details, such as undercuts, side cored holes or molded-in inserts. These extra cavity components add significantly to the mold's initial construction and ongoing maintenance costs.

2.10 Holes

Rotational molding is an ideal process for producing hollow seamless plastic parts. The very nature of the process makes it difficult to mold holes through the walls of a part. This is an advantage in the production of floats and balls. Most products require an opening for egress into the part, or for attaching the part to other things. Many holes are cut into parts after molding, but it is normally better to mold the hole in wherever practical.

Molders and tool makers have been resourceful in developing ways to provide holes in rotationally molded parts. For example, all of the blind and through holes shown in Fig. 16, except B², are routinely being produced.

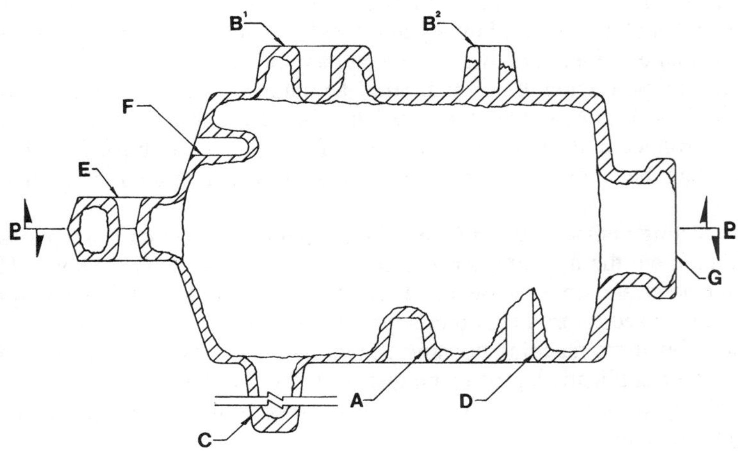

FIGURE 16 Rotationally moldable holes

The inwardly projecting blind holes, A and F, are produced on cores that extend into the cavity. These tooling cores must be designed to conduct enough heat to the free end of the core to ensure that they are coated with plastic material.

The outwardly projecting blind hole, or Boss B¹, can be molded because it is surrounded by open areas large enough to allow the material to flow properly. The outwardly projecting solid boss, B², cannot be made by rotational molding. This solid boss does not provide adequate open space for the flow of material.

The outwardly projecting hollow boss, C, can be molded into a cavity provided in the mold. The minimum diameter that can be molded is five times the nominal wall thickness of the part. If an outwardly projecting through hole is required, the cavity can be extended and the excess material can be cut off to open the hole.

Inwardly projecting through holes, such as that shown at D, can be produced with extended cores that are too large to allow the material to completely coat the core.

Most rotationally molded parts require a vent tube to equalize the pressure inside of the cavity during the heating and cooling portions of the molding cycle. These vent tubes can sometimes be used to provide holes of the type shown at D.

The through hole, E, is an inner column which is actually an open kiss-off rib. Through holes of this type are frequently used for mounting or attachment purposes.

All of the holes shown in Fig. 16 can be molded in a simple, two-piece mold, except for Hole F. This inwardly projecting blind hole is molded on a core which is parallel to the mold's parting line. This side acting core becomes an undercut. This core must be removed from the mold before the part can be demolded.

Large openings through the wall, as shown at G, can be produced by a unique method called "drop out sections" or "drop top parts". To form this type of opening, a section of the cavity is designed with a nonadhering surface on a thick mold section with low thermal conductivity. There is insufficient heat on the inside surface of the cavity in this area, and the plastic does not adhere to it, thereby forming the opening through the wall. Many open topped containers and drum liners are produced in this manner. Fig. 17 shows a mold and a molded part of this type.

One problem in producing holes with this technique is that the thick mold section, with its low thermal conductivity, shields the surrounding surfaces on the mold. This can produce undesirable thinning out of the part's wall beside the hole. This condition can be avoided by shielding an area which is smaller than the desired opening.

Another interesting approach to producing large openings is to mold two parts as one piece, and then cut them apart after molding to produce two separate parts. The refuse container body and cover shown in Fig. 18 is an example of this technique. This process is also used to produce other sophisticated shapes, such as the top and bottom halves of business machine housings, lawn mower grass catchers, and carrying cases. In other applications, the technique is used to produce two totally unrelated parts, such as two open topped planters, two x-ray film developing tanks, or two refuse container bodies.

This common approach to producing openings is called "head to head" molding. The process requires a secondary machining operation. However, it does have the advantage of producing two parts on the same molding cycle.

2.11 Corner Radii

The large corner radii that are a common feature on rotationally molded plastic parts have two primary functions. Stresses tend to be concentrated on corners. Sharp corners concentrate these stresses even more. Radiusing the corners helps distribute the stresses over a broader area to produce a stronger part.

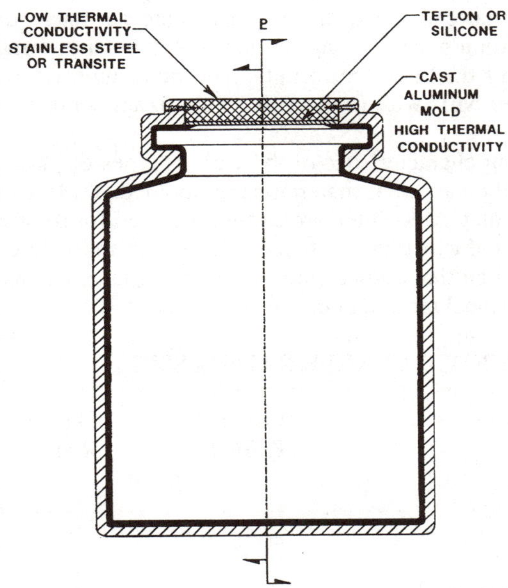

FIGURE 17 Large holes are produced by the drop out section method
Reproduced courtesy of the Association of Rotational Molders

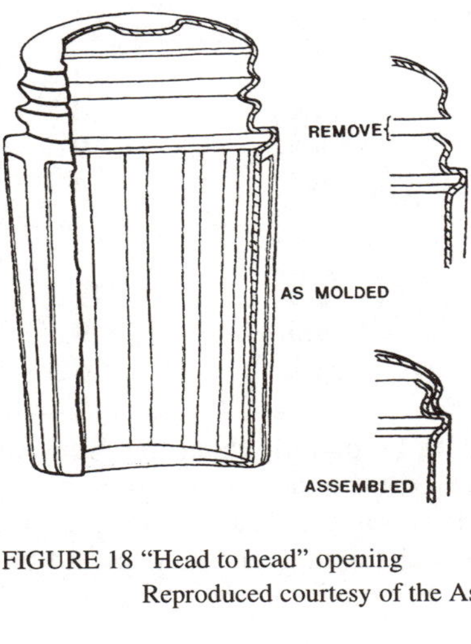

FIGURE 18 "Head to head" opening
Reproduced courtesy of the Association of Rotational Molders

Large corner radii improve the flow of the plastic in the cavity. Sharp outside corners tend to accumulate extra material. Sharp inside corners are normally thinner than the rest of the part. Providing generous radii on inside and outside corners produces parts with improved wall thickness uniformity. The more uniform wall thickness, in turn, produces molded parts with more uniform shrinkage with less of a tendency to warp.

Because of the differing molding characteristics of the various types of plastics, corner radii must be specified for the particular material being molded. Table 2 lists the recommended radii for plastic materials which are commonly used for rotational molding. These radii will be satisfactory in most cases. However, the nature of the rotational molding process is such that corner radii should be as large as possible within the confines of the functional requirements of the product.

Table 2. RECOMMENDED CORNER RADIUS SIZE

Material		Inside Radius (in mm.)	Outside Radius (in mm.)
Polyethylene			
	Ideal	12.5	6.5
	Commercial	6.5	3.2
	Minimum	3.2	1.5
Polyvinyl Chloride			
	Ideal	9.5	6.5
	Commercial	6.5	3.2
	Minimum	3.2	2.0
Nylon			
	Ideal	19.0	12.5
	Commercial	9.5	9.5
	Minimum	4.7	4.7
Polycarbonate			
	Ideal	12.5	19.0
	Commercial	9.5	9.5
	Minimum	3.2	6.5

The ideal shape for rotational molding would be one in which the radii are so large that the part becomes a round ball. Most parts cannot be designed as round balls. More complicated shapes often create shallow angles leading into the corners of a part. This condition creates closely spaced walls with a tendency toward bridging (Fig. 19).

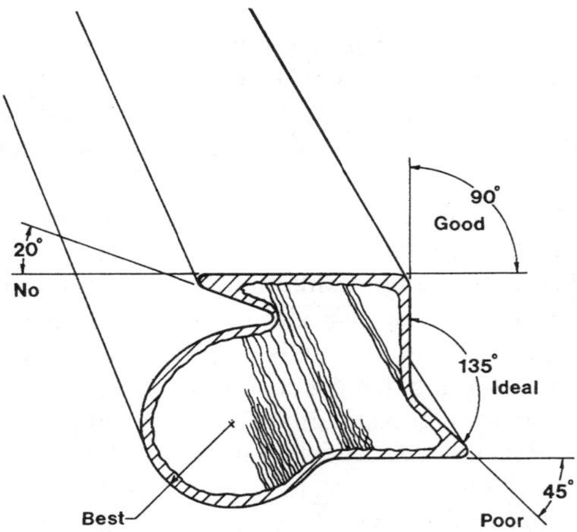

FIGURE 19 Corner angle limitations

The rotational molding process can easily produce parts with 90 degree angles in any material, as long as the corners are radiused.

Harder flow materials, such as polycarbonate, have difficulty in filling angles less than 45 degrees without bridging, which results in unfilled corners or solid sections.

The minimum recommended angle for polyethylene and polyvinyl chloride is 30 degrees. Corner angles which are as small as 20 degrees have been produced successfully in the easy flowing nylon materials.

The limiting factor on corner angles is that there must be an adequate radius at the point where the two walls meet. The design guidelines given for minimum wall separations and corner radii also apply to corner angles.

2.12 Tolerances

The important subject of which dimensional tolerances are possible for a given plastics processing technique is always difficult to reduce to hard, fast numbers. Attainable tolerances are a very individualistic thing that is dependent upon the design of the part, the plastic material specified, the dimensions of the cavity, and the processing technique which is used. Each combination of these four factors is an individual case which must be considered separately.

In finalizing a design, it must be remembered that small parts can be held to closer tolerances than large parts. Thin walled parts will be of more precision than the same size and shape of a part with a thicker wall.

It is generally agreed that plastic materials which have the smallest mold shrinkage factors will produce the more dimensionally stable parts. Amorphous materials, such as polyvinyl chloride and polycarbonate, are less susceptible to molding cycle variations than are crystalline materials such as high density polyethylene and nylon.

There are always some minor variations in the batch-to-batch uniformity of the plastic material as received from the material supplier. However, in more recent years, these variations are becoming less of a factor. Greater dimensional and processing variables can be encountered by using reprocessed, off-grade or improperly ground plastic materials.

To a greater extent, the ability of a molder to produce a part within drawing specification is dependent upon the plastic material being used and the design, construction and precision of the mold. In the case of multiple cavity molds, the dimensional uniformity of the molded parts must also allow for the dimensional variations among the various cavities. Theoretically, there is no limit to the level of precision that can be built into a cavity, assuming that the cost of the mold is of secondary importance. There is no substitute for a good quality, precision mold.

Rotational molding is a low pressure process, and the two halves of the mold do not require the high pressure clamping which is common to blow or injection molding. As a result, larger tolerances are required on part dimensions that are perpendicular to the mold's parting line. Routine mold maintenance and cleaning of the parting line surfaces will minimize these dimensional variations.

It is generally agreed that processing variables account for the majority of the part-to-part dimensional differences in rotational molding. The condition is improving, however, as the process is more clearly understood and as more sophisticated controls are applied to the various segments of the molding cycle.

Once the design has been set and the cavity built, these two factors should remain constant. That leaves only two variables: the batch-to-batch uniformity of the plastic material and the repetition of the molding cycle.

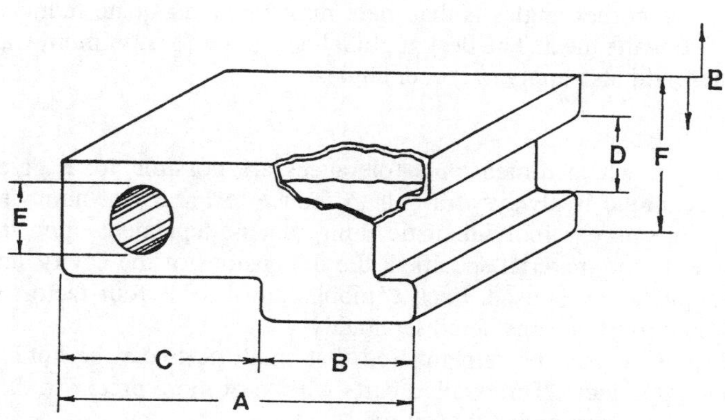

FIGURE 20 Dimensional tolerance considerations

If an exact dimensional reproduction is absolutely mandatory, the molder can devote all his skill and experience to that one requirement so that precision parts can be produced, but the resultant part cost may be prohibitive.

Care must be exercised in applying tolerances to part drawing dimensions, as over-specification will have a direct, adverse effect on the cost of the mold and the molded parts. As with all other plastics processing techniques, the best tolerance is the broadest tolerance that will satisfy the end use functional requirement of the part.

Table 3 can be used in conjunction with Fig. 20 as a design guide for applying dimensional tolerances to rotationally molded plastic parts. However, each part design and material selection is a special case which must be given individual consideration. The knowledge and experience of a custom processor can be very helpful in determining which tolerances will be practical for a specific part design.

The outside dimensions, Fig. 20 A, B, C and F, are free to draw away from the cavity as the plastic cools and shrinks. These dimensions will vary more than inside dimensions D and E, which are stabilized by the presence of the mold cores.

Dimensions and tolerances should not be applied to the inside surfaces of hollow rotationally molded parts. These freeformed surfaces are not in contact with the cavity, and they cannot be controlled with any repetitive degree of accuracy. In this regard, rotational molding is similar to blow molding and female thermoforming.

The design principles presented here are the basic guidelines that should be adhered to, regardless of the size or complexity of the part. As has been shown, these guidelines change slightly for different materials.

Table 3. DIMENSIONAL TOLERANCES

Material		A	B	C	D	E	F*
PE	Ideal	.020	.020	.020	.015	.010	.020
	Comm'l.	.010	.010	.010	.008	.008	.010
	Prec.	.005	.005	.005	.004	.004	.005
PVC	Ideal	.025	.025	.025	.015	.015	.025
	Comm'l.	.020	.020	.020	.010	.010	.020
	Prec.	.010	.010	.010	.005	.005	.010
Nylon	Ideal	.010	.010	.010	.008	.008	.010
	Comm'l.	.006	.006	.006	.005	.005	.006
	Prec.	.004	.004	.004	.003	.003	.004
PC	Ideal	.008	.008	.008	.005	.005	.008
	Comm'l.	.005	.005	.005	.003	.003	.005
	Prec.	.003	.003	.003	.002	.002	.003

NOTE: in ± cm/cm, plus cavity tolerance.
Ideal tolerance = minimum care required.
Commercial tolerance = possible with reasonable care.
Precision tolerance = possible with difficulty and added cost.
*plus .25mm for parting line variations

CHAPTER 8
Design Case Studies

G.W. Wassell

1. INTRODUCTION

Over the past decade the significant advances made in the field of rotational moulding have considerably widened the designer's scope. The process, and the advantages it offers, are now more widely known and there is a greater awareness of its true potential; this has resulted in an increasing demand for well designed rotationally moulded products.

When designers are commissioned for a project, this quite frequently involves replacing a product previously made in a more traditional material. In these instances it is important that the designer does not allow his thinking to be swayed by his client if suggestions made are not practical for the process. It is all important that the product is redesigned with the main aim of producing a satisfactory rotational moulding.

There is now a wider range of materials available, some "tailor made" for the process, and coupled with improvements in tooling, techniques etc, we are now seeing products designed - and produced - of quite a complex nature. These are a far cry from the relatively simple containers which were predominant, if not absolute, in the earlier days of rotational moulding. A good example of this can be appreciated from the "Garden Bar Trolley', shown in Figure 1. It was initially intended to manufacture the trolley as an injection moulding, but a successful redesign converted it to a rotational moulding. This resulted in the number of components being reduced from 8 to 3 and enabled the trolley to be marketed with a considerable reduction against the original estimated tooling costs.

2. CASE STUDIES

2.1 The FRIDO Playball

The first study relates to the birth and development of the FRIDO playball. This takes us back to 1954 when the late Mr Siegmund Friedland (who headed V & E Friedland Ltd and V & E Plastics Ltd) visited the annual toy fair in Nürnberg, Germany and saw a vinyl (pvc) playball on the stand of a Dr Priemer. He asked Dr Priemer to sell him the "know-how" of the ball manufacturer but this was refused,

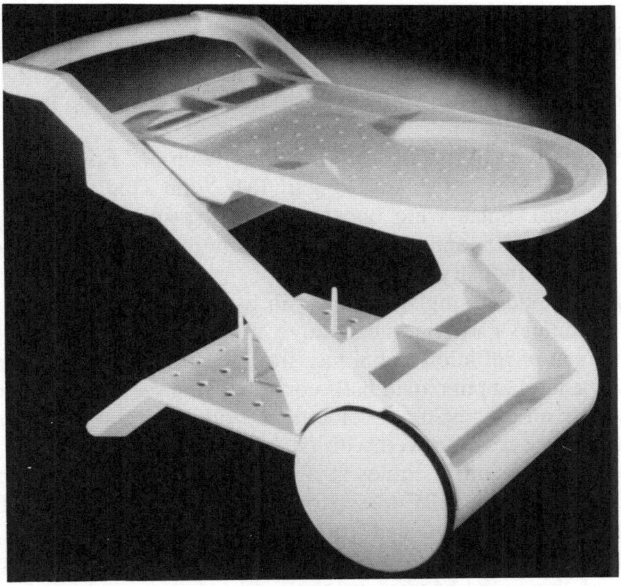

FIGURE 1 The Garden Bar Trolley; winner of the 'Overall product of the year' and 'Best conversion of a product from another process' awards at the 1990 Annual Meeting of the Association of Rotational Moulders
(photograph by permission of Lin Pac Rotational Mouldings)

despite Mr Friedland offering to disclose a technique for producing the "Tootsie bootie" – a lady's lightweight rain-bootee produced by hot-dip moulding. Mr Friedland returned to England and attempted to produce a ball, using plastisol in a modified domestic copperball float ... This he rotated by hand over a gas ring in the works canteen! Thus the idea was born.

A machine was built on which the moulds, containing pvc plastisol, gyrated in three planes of rotation, heated by direct gas flames. The heating "carried" the cure, or fusion, of the natural clear plastisol to a very slight degradation, so that the ideal cure (fusion) time was when the resultant moulding was a "touch" burned. The colour was the familiar amber-clear often seen by pvc technologists when studying heat stabiliser tests. With this degree of fusion, pvc exhibits maximum physical properties and thus maximum playball wear life. The familiar red colour of the well known 'A' ball was produced at a later stage.

Mr Friedland returned the following year to the Nürnberg Toy Fair and showed Dr Priemer the FRIDO ball. The difference in the two products was that the one produced by Dr Priemer was made in two halves by slush moulding, followed by bonding the halves together, whereas the FRIDO ball was seamless, made in one operation. This resulted in Dr Priemer purchasing FRIDO moulding machines, as did firms in France and South America!

Production of the balls commenced at the V & E Plastics works in Macclesfield, but demand became insatiable and in 1957 manufacture was moved to a factory purchased in Stockport, where 24 hour shift working was introduced. Plasticiser was delivered to the factory in bulk road tankers where in-house colouring was carried out on many mixing machines.

The name FRIDO was derived by taking the first three letters of Friedland and placing them before the first two letters of Doggart (Mr John Doggart being a director of V & E Friedland). Sales were boosted using the sponsorship of Sir Stanley Matthews of football fame, and Gus Risman for rugby. FRIDO became a household name in the playball industry.

Initially, the balls were moulded singly, but later in a cluster of five, or more, smaller moulds. These mouldings carried a metal valve, and the uniform wall thickness assisted the development of inflation techniques from the smaller mould size to the final ball size. It also meant that punctured balls could be repaired and reinflated.

Designs on the balls were an interesting development. Following the initial pimple or nodule surface, moulds were engraved with grooves which produced ribs on the ball. One was grooved in concentric circles at right angles so that the final ball looked like the model introducing the BBC television programme "Panorama". The signature "Stanley Matthews" was reproduced from an engraving in the mould.

Later, the Polka Dot ball was developed, where the surface spots were produced by placing disc shapes of thin pvc sheeting on the walls of the mould, prior to adding the ball-wall plastisol. The Rainbow ball was added to the range by spraying the moulded ball. Swirled, or marbled effects were achieved by painting the mould with high viscosity coloured pvc plastisol, prior to adding the main, wall coloured plastisol.

To reach higher production rates, Plastisol dispensers were made in-house, these discharging five measured shots of plastisol simultaneously into five moulds.

A number of patents and registered designs were granted as a result of the above work, for which it is right to mention the late Mr Jack Sutton who was so instrumental in making FRIDO such a success story.

2.2 Compost Bin

The second study depicts a COMPOST BIN, introduced in 1990 and shown in Figure 2. Although this is a relatively unsophisticated product, it has been selected to show how, nevertheless, certain design factors were applied to it.

Following a review of the horticultural market, with its increasing emphasis on organic gardening, it was felt there was scope for a functional, attractively designed compost bin.

Before commencing any actual design work, it is essential that the designer is able to consider all relevant factors. Two such factors which can have a bearing on the final product design are:

a) which section of a particular market is being aimed at - and what are the anticipated sales forecasts?

b) what is the loading capacity of the machine to be used in production?

FIGURE 2 Lin Pac Rotational Mouldings' Compost Bin, reproduced with permission

Only too often an insufficient design brief is given, perhaps because it is felt that factors such as above are "no business of the designers."

After considering a), it was decided to aim at the largest section of the market, ie the small to medium garden. This indicated the desirable bin capacity whilst the potential volume established that a maximum tooling complement was justified.

The shape needed to suit easy transportation home by the purchaser, and the initial configuration decided upon revealed that three units could be moulded per cycle on the planned production machine. By an adjustment to the initial product dimensions (and without any adverse effect on the cubic capacity) it was possible to increase the number of units per cycle to four. The significant impression this had on cost is obvious, particularly with a product selling through the chain of manufacturer - wholesaler - retailer - consumer.

It cannot be overemphasised how important it is for the designer to keep the cost aspect in mind. The relationship between Sales/Marketing and the Designer is all important and is a two way relationship.

The bin affords a good sized opening for ingress of the compostable materials, the lid being part of the body moulding - cut off by routing as a post-moulding operation.

An essential design feature was the need to provide a door at the base for easy removal of compost. This was achieved by providing guides in the moulding to take a double skin door. The guides being moulded in avoided add-on items and thus minimised post moulding work, see Figure 3.

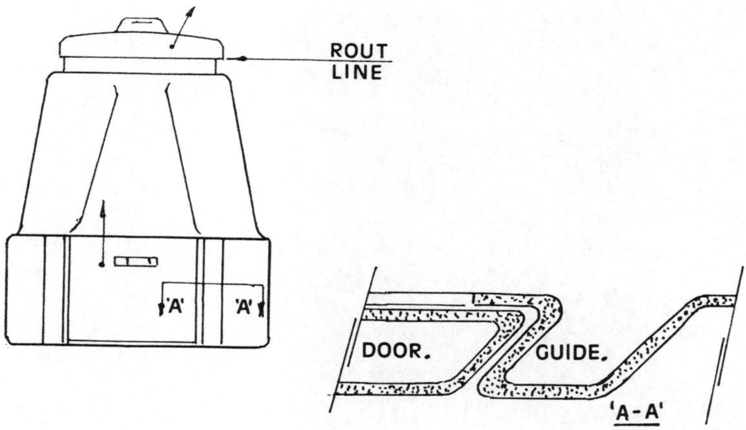

FIGURE 3

Rotational moulding designers are well advised to keep in mind the good old adage to "keep it simple" and not "design for design's sake". A complex design might, for example, result in a several part tool which can:

* increase machine operating times;
* increase mould parting lines and subsequent finishing operations;
* reduce the number of tools which can be utilised;
* inflate tooling costs.

The final compost bin design resulted in single part tooling, fully utilising the machine capacity available whilst the sheet metal tools ensured a minimum machine cycling time.

At a preliminary stage of any design a material section has to be made - in this case medium density polyethylene, dark green. During the actual composting period a limited amount of heat is generated, which is a "plus" in low temperature winter conditions, further confirming the decision that, in this application, reclaimed material (subject to certain controls) was perfectly suitable.

It is perhaps inevitable that the second case study should involve the use of polyethylene. Over the period between the two case studies polyethylene (in its differing modes) has become the most widely used material in the rotational moulding industry and it is this to which most of the design criteria apply.

3. ASPECTS OF DESIGN

As with all product design the basic essentials are that the product will be functional and aesthetically pleasing.

3.1 Product Design

During the product design, the subsequent tooling should be kept well in mind - to avoid problems at a later stage. The tooling to be used may well have a bearing on the product design. The main types used are:

Fabricated (steel, aluminium, stainless steel)
Cast aluminium
Electro formed

In this chapter it is not intended to cover tooling construction in detail, although it is inevitable that some aspects are touched upon. Certainly, product designers need a working appreciation of each type of tooling manufacture. There are a number of factors which will influence the final decision on tooling including, of course, that of economics. Unquestionably, however: the better the tooling - the better the end product!

Without going into all the advantages/disadvantages of each type of tooling, it is probably fair to say that:

- ✱ Fabricated tooling will almost certainly be selected for larger tools and where the number off is limited;
- ✱ Cast aluminium is likely to be the choice when the shape is complex and involves several or multi-tooling;
- ✱ Electro formed is for precision items involving fine detail and special surface finishes such as leather graining.

The surface finish on any moulding has a very real impact on its appeal; a gloss surface may be preferred and this will entail the tooling being highly polished. Conversely, a matt finish may be chosen, and again this needs specifying at the tooling making stage, in which case the tooling is normally shot-peened or aquablasted. A combination of these surface finishes can be attractive; an example of this is the lawn mower box shown in Figure 4. This moulding is produced from 3 part cast aluminium tools, the main panel being matt, with the side panels having the gloss finish.

Logos, or other details can be reproduced from an engraving into the tool, letters as small as 3mm in height being well defined in the finished moulding.

3.2 Materials

Materials are covered in other chapters, and the following brief coverage is made only in its relation to the product design. The product end-use and the environment in which it will function are obviously the overriding factors when considering the material selection. Whilst polyethylene (available tailor-made in varied grades for a wide range of rotational moulding applications) is the most commercially used, other materials available include polypropylene, polycarbonate, pvc, nylon. The melt flow and grind of particle size have a bearing on the flow of material into

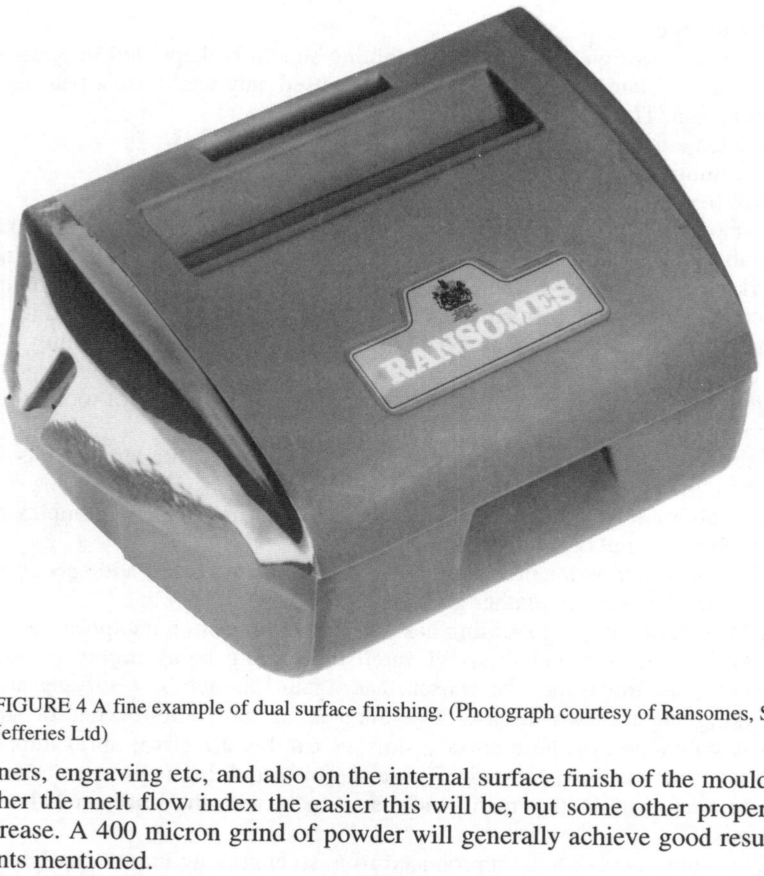

FIGURE 4 A fine example of dual surface finishing. (Photograph courtesy of Ransomes, Sims and Jefferies Ltd)

corners, engraving etc, and also on the internal surface finish of the moulding. The higher the melt flow index the easier this will be, but some other properties may decrease. A 400 micron grind of powder will generally achieve good results in the points mentioned.

Cross-linked polyethylene is particularly noted for high stiffness and resistance to low temperature. Polypropylene can also provide particularly good strength and stiffness and remain stable at high temperatures - but can embrittle in low temperatures. Because of its increased stiffness, it is important that tooling does not include any undercuts which can make de-moulding difficult, whereas in the same condition it may be possible to "spring" a polyethylene moulding from the tool.

When coloured material is to be used in a product to be exposed to sunlight, it is important that it carries a stabiliser to resist degradation. In an outside environment, compounded material is preferred to the dry-blending-in of the colouring pigment in terms of colour fastness and mechanical strength.

Other additives used include, for example, flame retardants and anti-static agents.

3.3 Distortion and Shrinkage

Distortion and shrinkage are possibly two of the most common problems which the rotational moulder has to consider in endeavouring to ensure a perfectly acceptable and aesthetic product. High density material has a greater tendency to distort than low density and also a higher degree of shrinkage. Whilst both these problems can, to some degree, be tackled in production (principally by the method and rate of cooling and ensuring adequate mould venting), no amount of care will compensate for a badly designed product.

Large flat surfaces are most prone to distort and it is useful to keep in mind the surface finish to be specified, since a matt or textured surface can help disguise distortion by breaking reflected light, whereas a gloss surface will emphasise any distortion.

3.4 Ribbing

The most positive way of dealing with distortion at the design stage is to design some "form" or change of shape into the product - or by stiffening ribs or swages.

The industrial container (size 1.2m x 1.0m x 1.0m) shown in Figure 5 is a good example of the use of ribbing, not only to obviate distortion, but also to add structural strength, in this instance allowing the containers to be stacked. The ribs in the base provided a means of handling by fork lift truck, and these were given a generous draft angle of 10° to assist demoulding.

Typical ribs/swages are shown in Figure 6. The type 6a should have a draft angle of 2° to 3°. Its width must be adequate to avoid "bridging" of material and not too deep to cause heat restriction and consequent material thinning.

It is not possible to dictate the sizing or spacing of ribs since this depends on many factors. The radii on ribbing should follow the recommendations covered in that topic.

Well designed ribbing on, say, an industrial container, will obviously mean an increased surface area, but the additional rigidity and strength it provides will almost certainly mean a saving on the raw material shot weight.

Another type of ribbing or coning is shown in Figure 7. This type of support is frequently used where there are two walls, say up to 75mm apart, which it is necessary to stiffen to provide a greater load bearing. The most common application is in industrial pallets and in the base of twin walled containers. Where ribbing as in 7b is chosen, the length of the ribs must not restrict the free flow of material during the heating cycle and this can be assisted by staggering the ribbing.

The gap to be allowed in the tooling between the cone tip(s) and inside of the opposing tool face is rather critical and is normally approximately 1¾ x wall thickness. It is often found in practice that after prototype mouldings have been provided, there is a desire to either increase or decrease the planned wall thickness. It is, in the case of cones, wise to consider having the tips made replaceable so that the height can be adjusted in the event of an alteration in the shot weight.

FIGURE 5 (Photograph by permission of Lin Pac Rotational Mouldings)

3.5 Clamping

There are, of course, instances where ribbing is impractical, is not needed for structural strength, or is undesirable aesthetically. In such cases clamping into a wooden former immediately after demoulding can help to contain distortion. Although the initial cooling time is critical, it has to be remembered that the total shrinkage period may be up to 24 hours, and to be effective in production a number of the formers will therefore be necessary.

The shrinkage allowance in the tooling design can vary depending on, for example, the material, product configuration, and production. Material such as polypropylene has a much lower shrinkage factor than, say, polyethylene for which a shrinkage of 2 - 3.5% is normally calculated. Where final dimension tolerances are tight, again it may be possible to use wooden formers to good effect.

3.6 Corner Radii

In design, attention to this aspect is of the utmost importance. An internal radius on the tool will tend to give an increase in the product wall thickness (Figure 8a), and so add to its strength. This increase in the nominal wall thickness will gradually

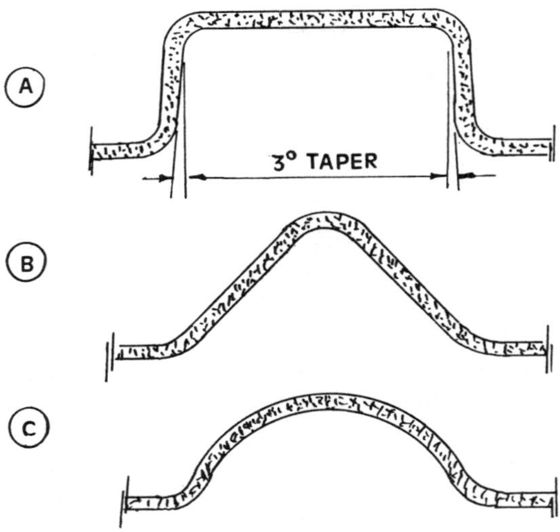

FIGURE 6

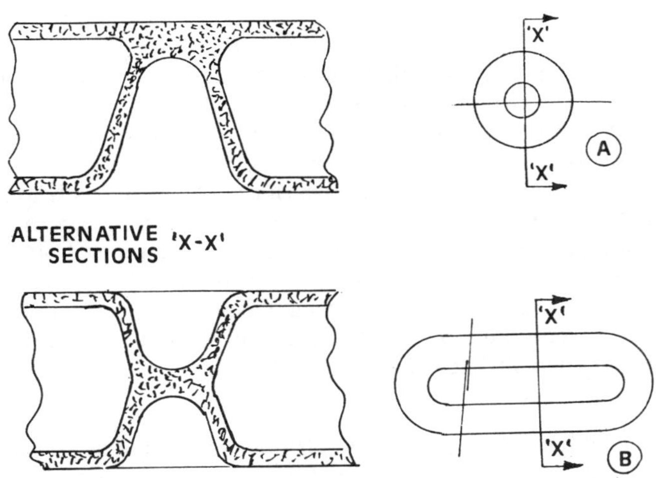

FIGURE 7

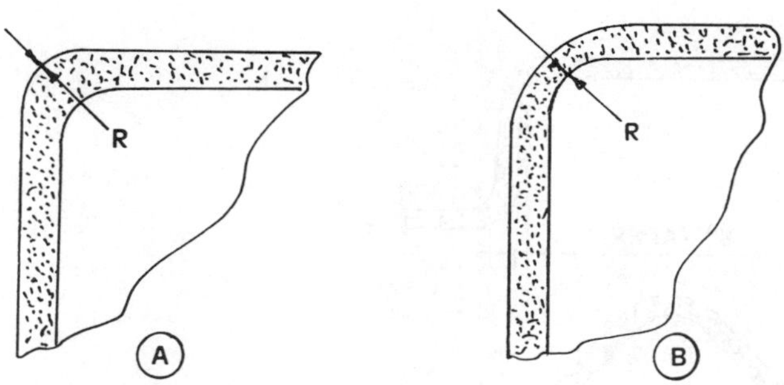

FIGURE 8

decrease as the corner radius increases, to give a thickness consistent with the walls. A square corner is to be avoided, since material will not flow into it fully, and will result in a "perforated" effect.

On an external tool radius (internal product radius), the material will, if the radius is too tight, tend to thin out around the radius. This is undesirable (Figure 8b). Therefore, a radius of less than, say, 6mm on polyethylene (10mm on nylon) is not recommended.

An ideal rotational moulding is one where a smooth, flowing contour appertains and this can be best achieved by incorporating the maximum radii possible.

It is also good practice, when possible, to keep a uniformity in radii specified. When differing radii come together at intersections or corners, it complicates tooling manufacture and can adversely affect tool costs.

3.7 Part Lines

Many otherwise excellent rotational mouldings have their appearance spoiled by insufficient attention being paid to the mould parting line(s). The ideal is to "design out" these altogether if possible, and this can sometimes be done, on, for example, containers having an open top.

When the parting line is visible on the moulded part it can be labour costly to try to achieve a satisfactory appearance. One method used as a means of concealing the parting line, is to emphasise it! This can be done by bringing it on to a bead in the moulding (Figure 9) which makes it far easier to remove any likely joint-line flash without causing surface damage to the moulding. Good joint faces with adequate clamping and mould vents are tooling essentials relating to this aspect, as is care in subsequent production to obviate abuse of joint flanges.

3.8 Wall Thickness

It has been previously mentioned that an attribute of rotational moulding is that a good uniform nominal wall thickness can be achieved. Provided the product has been carefully designed and production correctly carried out (machine rotation speeds/ratios can have a bearing) then this fact is true.

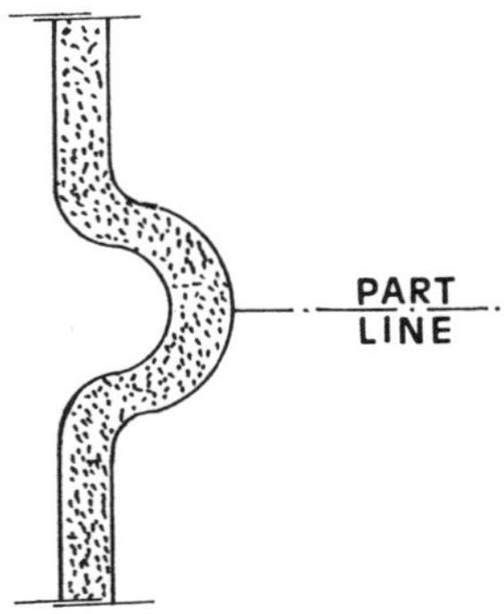

FIGURE 9

One aspect that can prevent uniformity in the wall thickness is insufficient heat penetration during the heating cycle. This can be the case where, for example, the joint mould flanges are too heavy. On, say, a fabricated mould these should not exceed a thickness of approximately 10mm – the two flanges together giving a thickness of 20mm. When, possibly due to the tool configuration, these result in heat restriction, the mass can be reduced by drilling a series of close pitched holes along the flanges. Typical flanges are shown in Figure 10.

When detailing the machine mounting work where a number of moulds are involved, they must have sufficient space between them to allow heat to circulate freely. Figure 11 shows an unsatisfactory mounting.

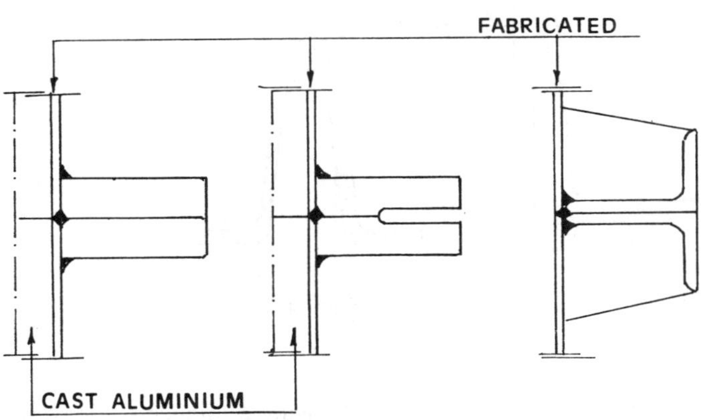

FIGURE 10

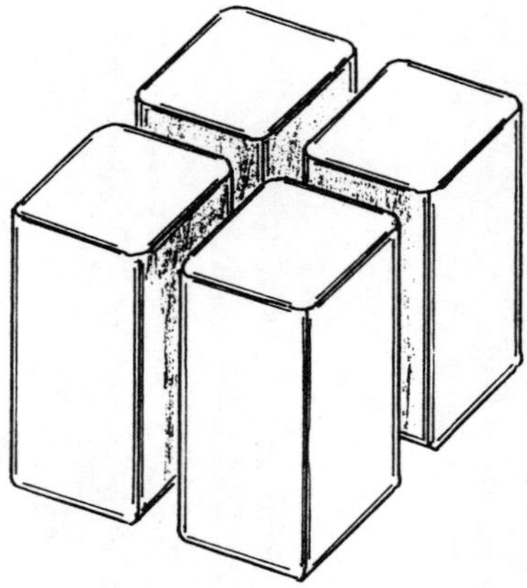

FIGURE 11

Figure 12a illustrates a double walled container. Sometimes insufficient heat gets down into the inner section of the tool to achieve a satisfactory thickness of the inner moulding walls. This can similarly happen where there are deep pockets in a moulding (Figure 12b).

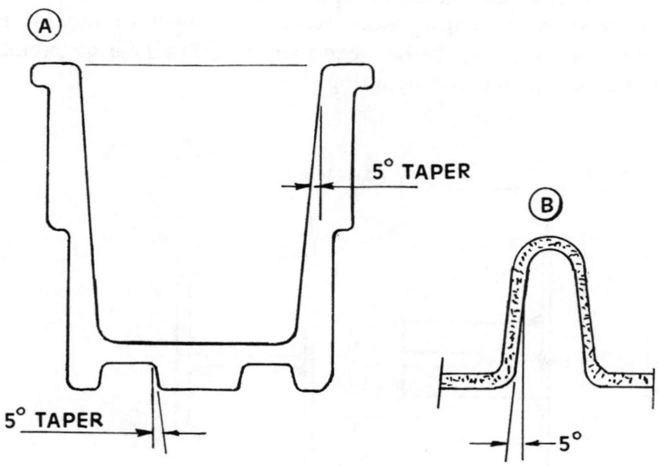

FIGURE 12

In these cases it is advisable to design the tools using a thinner section for the areas which can cause difficulty, with a thicker section for the remainder of the tool. An alternative is to use materials with a differing thermal conductivity.

Where cast tools are used, the outside can be machined down to create a variance in tool wall thickness. (Any such machining must be carried out on the outside surfaces, since machining may possibly disclose any porosity present beneath the surface which would be unacceptable on the inside.)

Another way of increasing the wall thickness in one area of a moulding is to fix a "heat shield" made from wire mesh or expanded metal over the remaining area of the tool. The resulting imbalance of heat input enables the unshielded area to build up a heavier wall section.

3.9 Double Walled Mouldings

With this type of moulding the main factors to be considered are:

a) That the cubic capacity of the cavity is large enough to take the relevant shot weight of material. (In extreme cases it may be necessary to design the tool so that an initial charge can be partly fused - and then a second charge applied.)

b) The width between the tool faces is sufficient to enable material to flow freely and not "bridge". The ideal is 5 x wall thickness.

c) In the case of a container such as in Figure 12a, or pockets as in 12b:

   i) since material will shrink on to the inner tool section, this must have draft angle of approximately 5°.

   ii) to ensure that sufficient heat reaches down into the inner tool section (see advice under Wall Thickness).

   iii) de-moulding this type of container can be difficult. A lip around the top of the outer moulding section will assist.

4. INDUSTRIAL CONTAINERS

Industrial containers with an open top are routine rotational mouldings. With these, the amount of scrap material is reduced by heat insulating the top face of the tool (Figure 13a). The insulation must not extend to the edges - otherwise it can adversely affect the adjacent wall thickness. Venting is shown through the insulated area.

Assuming that the production machine is large enough to take a "double mould" (Figure 13b), then obviously scrap material is minimal and better machine utilisation is achieved. In this situation, venting can be through the line of the joint flanges.

4.1 Venting

Vents are fitted to relieve pressure build-up in the moulding, and should be adequate for this purpose. Assume the container in Figure 13 had a size circa 1m x 0.8m x 0.8m, then two vents of say 20mm diameter would be ideal.

A good material to use for the vents is PTFE tubing, and these should reach as far as possible to the centre of the mould. Where practical, they should be arranged to enter through a scrap section of the moulding - or where a cut out in the moulding is subsequently to be carried out.

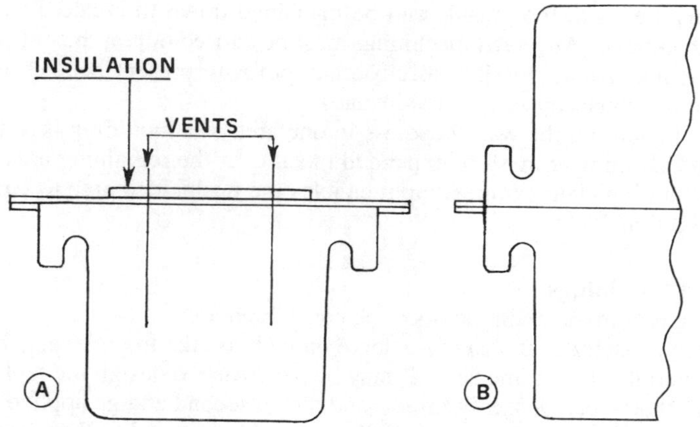

FIGURE 13

When water cooling is used to reduce the moulding cycle time, ingress of water through the vent can, occasionally, cause a "water-marking" on the inside surface of the moulding. A simple method sometimes used to reduce this is to apply steel wool into the outer end of the vent tube. Obviously, if cooling is restricted to air only, then the problem is obviated completely.

4.2 Inserts

These can be "moulded in" and a typical male and female insert are shown in Figure 14.

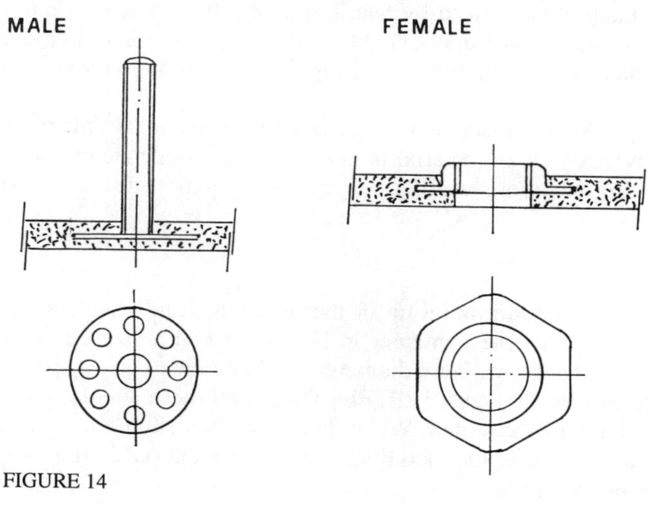

FIGURE 14

Metal inserts are normally made in brass, aluminium, or steel, those with a better thermal conductivity being preferred. They must be of thin enough section to ensure heat penetration and should not have sharp edges. There must be a means of holding the insert in the tool, to make certain it is located in the correct position relative to the wall thickness of the moulding. Whilst the moulding will shrink onto the insert, it should be made so that it is "locked" into the material. (The male insert shown has a perforated disc whilst the female insert has been knurled around the stem.)

When designing a product with inserts, thought must be given to its effect on the tooling relative to de-moulding, since several inserts (on differing faces) will almost certainly increase the number of split lines in the tool.

Moulded-in inserts should only be used for smaller diameter thread sizes; on larger sizes, consideration should be given to moulding-in the required thread form. As in other aspects, the thread will be better formed in one material than in another - but it is advisable to restrict moulded-in threads to a coarse or square thread type. It will be appreciated that with moulded-in threads a retractable core or collar has to be designed into the tool, to facilitate moulding withdrawal.

4.3 Undercuts

Where undercuts are essential in the product design, then careful thought needs to be applied to the placing of the parting line(s) in the tool. Undercuts which are parallel to the parting line can cause problems.

Since it is desirable to have the minimum number of part lines, it is sometimes possible with the more flexible materials to "spring" the moulded part from the tool. This, however, is not a practice to be encouraged since it can lead to surface damage and/or some part deformation.

In cases where the depth of undercut is slight (perhaps an engraving in the tool to a depth of say, 1.5mm) then in the case of the higher shrink materials, it should not create any difficulty. This class of material also has a bearing on any draft angles that are necessary.

4.4. Draft Angles

Assume the moulding is a simple box as Figure 15a, then in most cases the initial shrinkage of the material will enable withdrawal to be made even without any draft angle. Nevertheless, if aesthetically there is no objection to a draft angle being incorporated then this will always facilitate de-moulding (Figures 15b and 15c). When large, and particularly deep, mouldings are involved then one is certainly advisable, in the order of 3 - 5°.

4.5 Decoration and Logos

Previous mention has been made of engraving a tool to reproduce, say, a company logo. The use of dry-fix transfers/decals (as a post moulding operation) is a way of adding attraction and "life" to a product (see Figure 16).

A more recent and increasing development is the use of moulded-in graphics.

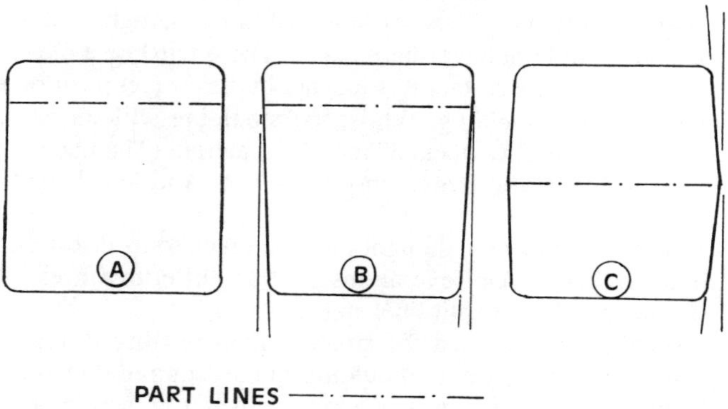

FIGURE 15

FIGURE 16 (by permission of Lin Pac Rotational Mouldings)

5. CONCLUSION

Inevitably, a significant content of this chapter will be accepted practice to a number of rotational moulders. For designers who have not had first hand experience of the process, the broad guidelines as set out are based on many years' practical experience. Coupled with the advantages (and limitations) of rotational moulding set out elsewhere, it should reaffirm the versatility of this manufacturing method.

Figure 17 shows a section of a pallet (approximately 1m x 1.2m) designed to handle a component during its movement, within the automotive industry. The unit is produced from a two part cast aluminium tool of comparatively low cost which, measured against the volume, made it another ideal application for rotational moulding.

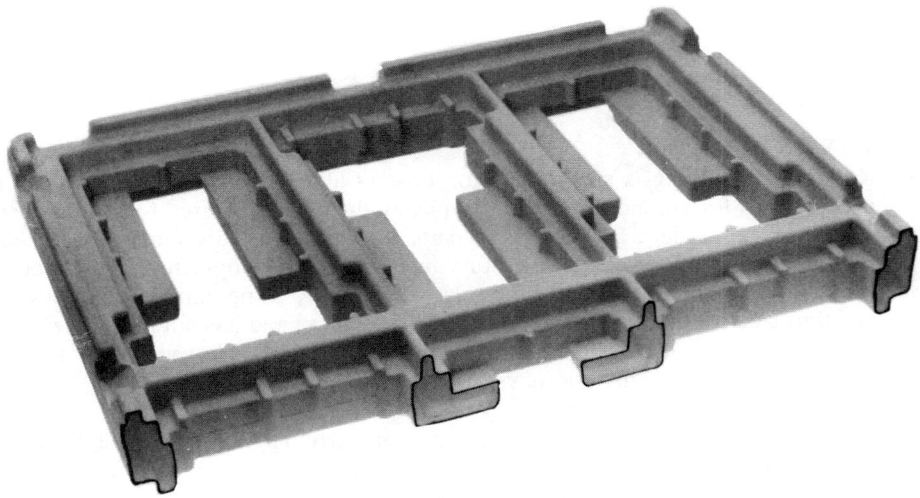

FIGURE 17 Pallet section (by permission of Lin Pac Rotational Mouldings)

ACKNOWLEDGEMENTS
To Mr Harry Grimshaw FPRI for information on the FRIDO playball case study.

The writer of this chapter would like to take this ideal opportunity of placing on record grateful thanks to those material suppliers, toolmakers and associates, who, over many years, have given immeasurable help and assistance.

CHAPTER 9
Process Control for Rotational Moulding

P.J. Nugent and R.J. Crawford

1. INTRODUCTION

Rotational moulding is still a fledgling process although it has been in use in its present form for around forty years. Typically the process is used for its simplicity and ability to produce low cost large parts economically. This simplicity and ease of use has however been a millstone in terms of development of the process. Moulders are able to produce parts relatively quickly using experienced operators and thus true control of the process from inside the mould has not been a necessity for relatively low-tech and non-demanding applications. Nowadays the trend is towards more technically and physically demanding parts with resulting pressures on moulders to provide better quality control.

As a result the machinery and techniques of rotational moulding have developed only in terms of functionality and engineering quality. By comparison with other processes such as injection and blow moulding, the levels of instrumentation and process monitoring in rotational moulding are minimal. Until recently, closed loop control has only been the domain of blow and injection moulding.

2. PRINCIPLES OF PROCESS CONTROL

If temperatures are recorded at various positions in the mould of a typical carousel type rotational moulding machine, then the graphs shown in Fig. 1 are obtained. In this case the set oven temperature is 300°C. The upper line shows the temperature rise of the air inside the oven once the mould has entered the oven and the doors have been closed. This temperature tends towards the set temperature and its proximity to this will depend on where the oven temperature is measured by the machine, the thermal efficiency and the oven time. It is interesting to note that most modern rotational moulding machines use the air temperature in the oven as the basis for feedback control of the process. However, it may be seen from Fig. 1 that this temperature shows little correlation with the temperature of the mould or its contents as indicated by the lines X, Y and Z. The latter temperatures are not independent of the oven air temperature or set temperature but they are influenced by other factors such as the amount and nature of the plastic in the mould, the surface area to volume ratio of the mould and the material and thickness of the mould

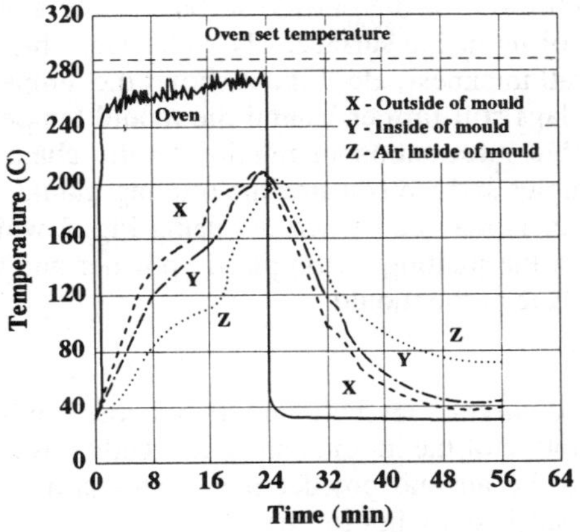

FIGURE 1 Temperature measurements in oven and mould during rotational moulding

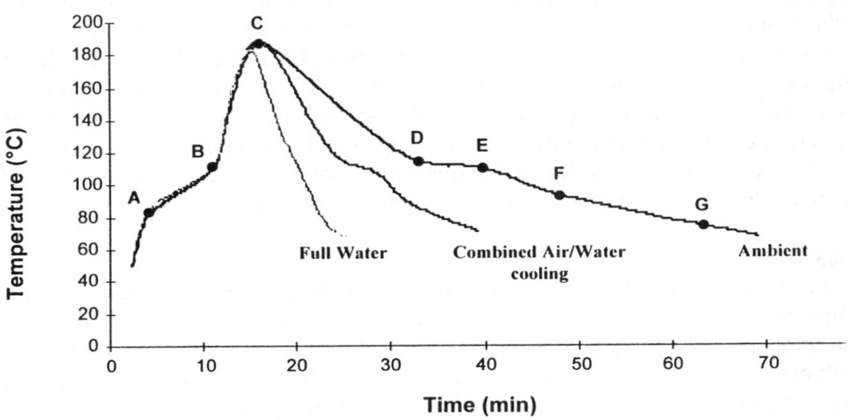

FIGURE 2 Typical internal mould air temperature profiles during rotational moulding

itself. As a result it is not realistic to expect to get precise process control based solely on the oven air temperature.

Line X in Fig. 1 shows the temperature rise of the outside surface of the mould and line Y shows the temperature rise of its inside surface. The latter lags slightly depending on the mould material, its wall thickness, etc. Line Z shows the temperature of the air inside the mould which lags still further behind the mould temperature but this line provides very valuable information in relation to the chain of events happening to the plastic as it tumbles and melts inside the rotating mould.

Closer examination of the line Z (Fig. 1) may be carried out using Fig. 2 which illustrates the various stages involved in the melting of the plastic powder and the formation of a coating on the inside surface of the mould.

Stage 1:
At the start of processing, the mould and polymer are cold and the powder tumbles about freely. Up to point 'A' the temperature of the internal air rises steadily as heat passes through the mould directly into the air and powder mass. This continues until the powder and mould surface are sufficiently hot for material to stick.

Stage 2:
At point 'A' the first layers of plastic adhere to the mould. The rate of temperature increase slows to form a plateau as successive layers of powder are laid up on the inner surface of the mould. The reduction of temperature rise of the internal air is due to energy absorption by the melting processes. Heat passing through the mould must pass through a layer of melting plastic; energy is absorbed and the rate of heat transfer is slower due to the lower thermal properties of the plastic.

Stage 3:
At point 'B' all the powder has adhered to form a molten skin with a powdery inner surface; the internal air temperature rises sharply once more as no more energy passing through the mould into the part is required for melting. After 'B' the temperature continues to rise as the melt consolidates and any trapped bubbles dissolve. The inner surface becomes smooth and melt density rises. The maximum temperature achieved inside the mould (Point 'C') is dictated by oven temperature and the time spent therein. Impact studies have shown that this peak temperature is directly related to the strength of the moulded product.

Stage 4:
When the mould moves into the cooling bay (just before point 'C'), the internal air temperature will rise slightly due to residual heat in the mould and molten polymer. The temperature then falls at a rate determined by the cooling medium employed. Ambient cooling is slowest, forced air cooling is faster and water cooling is the fastest. If the cooling rate is too fast it can cause warpage problems, whereas if it is too slow the process is not economical. In this stage the plastic is in the form of a hot melted layer on the inner surface of the mould, ie: it has not yet solidified.

Stage 5:
For some materials, such as polyethylene, there is a kink corresponding to the crystallisation point of the material (point 'D'). Latent heat absorbed during melting is released and this maintains the internal air temperature at a constant level; the length of the plateau is directly related to the volume of material in the part. Polyethylene shows this kink very well. The position of this second plateau relative to the melting region (A-B) can be linked with degradation due to the fact that degraded material will crystallise at a lower temperature.

Stage 6:
At point 'E' there is another inflexion in the curve and this is linked with the end of solidification across the wall thickness of the moulding. The material then continues to cool, and close inspection of the trace shows a slight change in slope at point 'F'. This is linked with the plastic moulding pulling away from the mould wall. It results in the plastic and the internal air cooling less quickly because they become insulated from the cooled mould by an air (or vacuum) gap. The points D, E and F can be identified quite clearly if the difference in temperature between the inner air and the inner surface of the plastic is recorded as shown in Fig. 3.

The process of cooling continues until the part can be handled or removed from the mould. The rate of cooling is less critical to the quality of the moulding in these latter stages but, as indicated above, rapid cooling can be difficult if the part has separated from the mould. Point 'G' on Fig. 3 shows a typical de-moulding point.

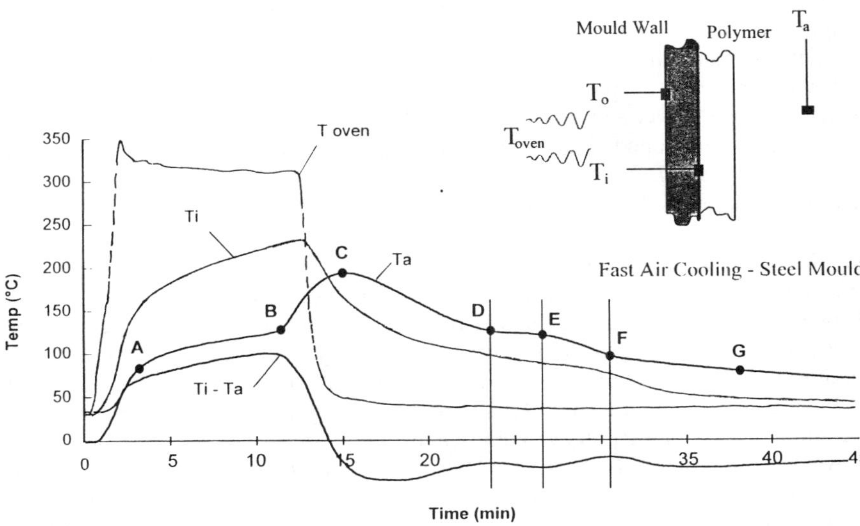

FIGURE 3 Internal air temperature and mould temperature during rotational moulding

The shape of the internal air temperature profile has been shown (2) to be consistent across a wide range of moulding conditions including mould material, mould size, mould thickness, part thickness, part material and oven temperature. The salient features remain the same with only the relative positions moving according to changes in conditions. It thus represents a consistent basis for quality control during rotational moulding. Measuring equipment such as Rotolog (3) is commercially available to record the required information.

2.1 Effect of Mould Material

Figure 4 shows the effect of variations in the mould material. Differences in thermal properties and indeed wall thickness will have marked effects on the cycle times and curing of the polymer powder. The two most commonly used materials for rotational moulds are steel and aluminium. Aluminium moulds tend to be thicker to compensate for lower strength properties. However, aluminium has better thermal transfer properties than steel so that the overall loss in thermal conduction is lower than for an equivalent thickness of steel.

During moulding, the temperature gradient across the wall of an aluminium mould will be shallower than for a thinner steel mould. This lower temperature gradient can cause problems when moulding materials such as polycarbonate which require rapid heating and high input rates. By being able to measure the temperature differences it is possible to assess the viability of a particular mould or mould material for a part. Balancing of different mould types is discussed in Section 4.1.

2.2 Effect of Part Thickness

Figure 5 shows how the internal air temperature is affected by increasing shot weight. For thin parts, the melting and cooling plateaus are shallow due to the small amounts of material involved. As the wall thickness increases, it takes longer to melt the larger bulk of powder and so the temperature plateau during which melting occurs becomes longer and flatter. For very thick parts (in excess of 20mm) the plateau can be almost horizontal. The temperature rises very slowly to just below the melting point of the material. As shown in Figure 5, the point at which the powder disappears can easily be seen as the transition becomes more dramatic for thicker parts.

2.3 Materials Other than Polyethylene

The internal air temperature profiles shown in Figs 1-5 are not unique to polyethylene. A range of other rotomouldable plastics exhibit similar behaviour. This includes: polypropylene (PP) (Figure 6), polycarbonate (PC) (Figure 7), PVC (Figure 8), ETFE (Figure 9), PFA (Figure 10), Hytrel (Figure 11), ABS (Figure 12), Nylon (Figure 13). The materials all exhibit a characteristic profile which differs from material to material according to the melting point and physical nature of each.

Amorphous materials such as Hytrel and polycarbonate do not exhibit a secondary cooling kink due to crystallisation, although both demonstrate a plateau during the lay-up stage. Crystalline materials such as polyethylene and poly-

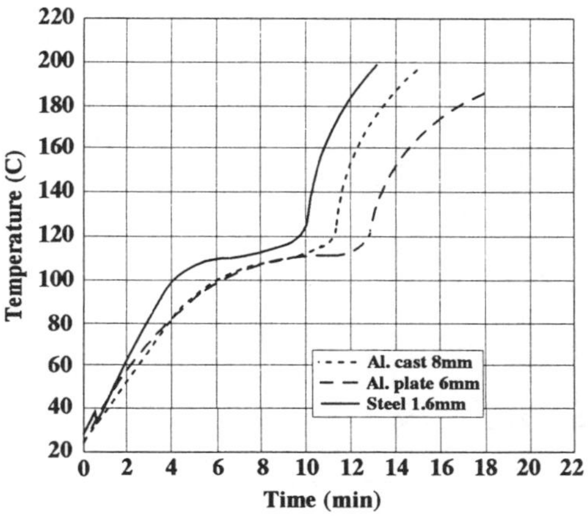

FIGURE 4 Effect of mould material on heating times

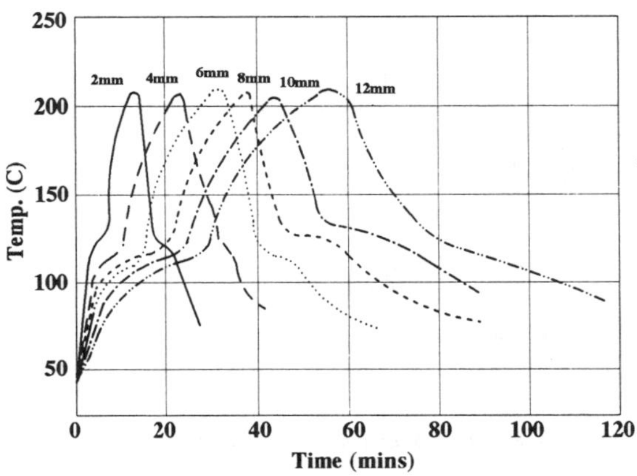

FIGURE 5 Effect of part wall thickness on mould internal air temperatures

propylene both release latent heat of crystallisation during cooling. The curve for polypropylene is very similar to that for polyethylene except that for PP the temperatures at which the various stages of transition occur are elevated.

High temperature materials such as ETFE and PFA require very high mould temperatures before the material begins to adhere. These fluoropolymers are commonly used for 'rotolining', ie: adding a thin barrier layer to the inside of a tank or vessel. In these cases the thermal performance of the container will be

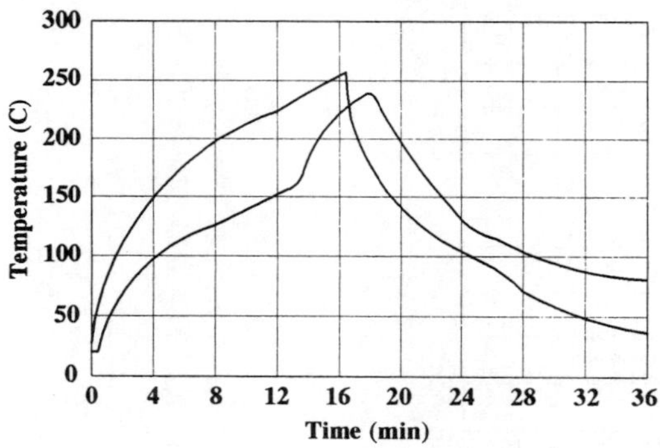

FIGURE 6 External mould wall and internal air temperature profiles during moulding of polypropylene (3mm)

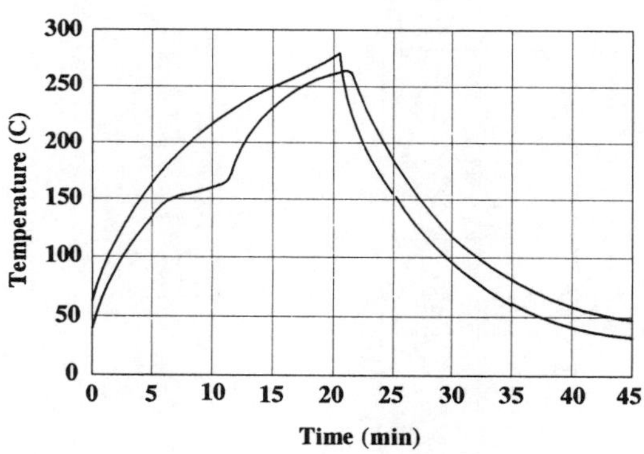

FIGURE 7 External mould wall and internal air temperature profiles for Makrolon polycarbonate (3mm)

unknown. It is very advantageous, therefore, to be able to watch the process as it occurs within the part. This is also important due to the high cost of fluoropolymers - errors can be very expensive.

Liquid materials such as PVC and Nylon behave differently due to the very different processes of lay-up on the mould surface. They differ from powder systems in the way in which they build up on the mould surface and in some cases the generation of an exotherm due to chemical reaction. Nevertheless, the air temperature curves do exhibit significant features and the maximum temperature experienced by the material can be used as a guide to the state of cure of the material.

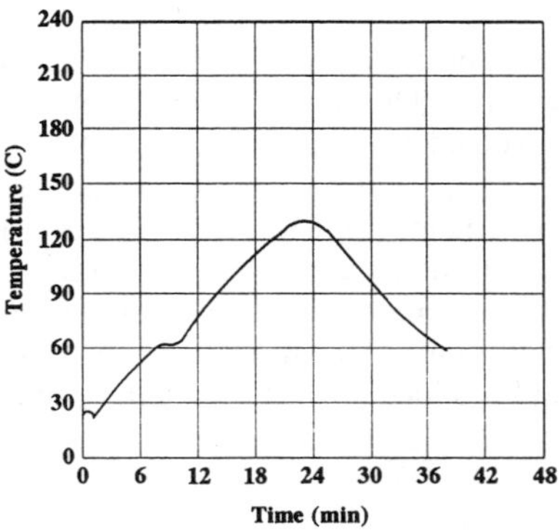

FIGURE 8 Internal air temperature profile for 5mm thick PVC cube moulding

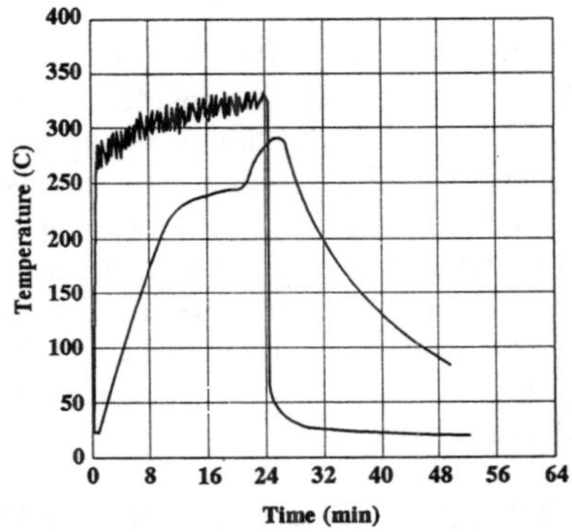

FIGURE 9 Oven and internal air temperature profiles for Hyflon (ETFE)Lolymer material (4.5mm)

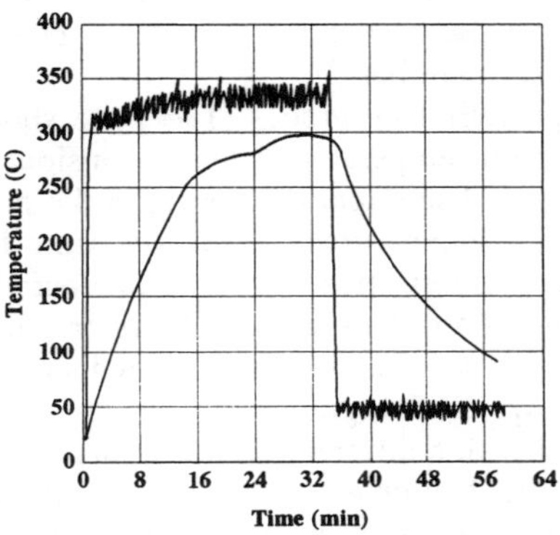

FIGURE 10 Oven and internal air temperature profiles for Du Pont PFA fluoropolymer material (3mm)

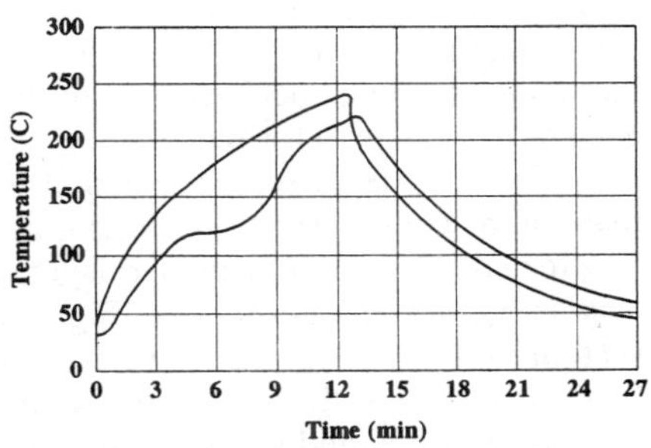

FIGURE 11 External mould wall and internal air temperature profiles for Hytrel 4056 material (3mm)

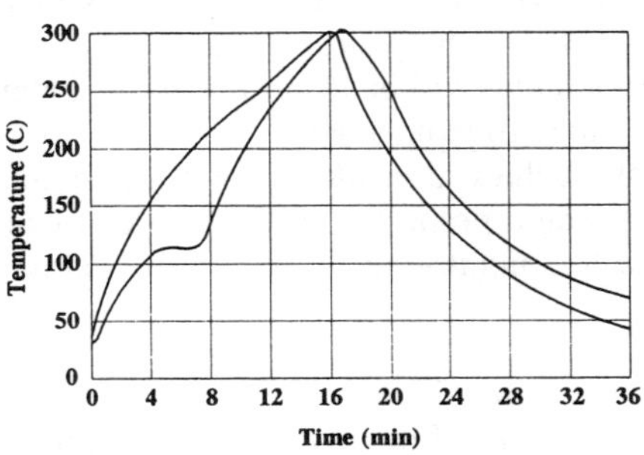

FIGURE 12 External mould wall and internal air temperature profiles for Ronfalin ABS (3mm)

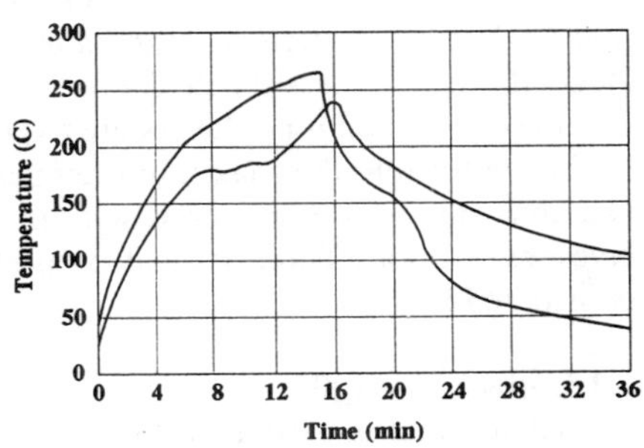

FIGURE 13 External mould wall and internal air temperature profiles for Capron Nylon (3mm)

In the past, an understanding of the processability of any new material could only be determined by considerable trial and error. This was time-consuming and expensive in terms of labour, equipment wastage and material wastage. Not only do changes in the material type affect machine settings but also physical changes in standard feedstock can alter the processability. Changes in Melt Flow Index (MFI) and powder flow properties can also affect cycle times, especially for large parts. By using the type of internal air temperature measurement described in this chapter, this trial and error approach is no longer necessary. The moulder now has the opportunity to control the process precisely and produce consistent quality products with confidence.

3. CONTROL OF PART PROPERTIES

The properties of the moulded product are generally crucial to any moulding process. The performance of a part when in service can be severely affected by its processing history, and material can be wasted when increased wall thickness has to be used to compensate for below strength material. Better control of material properties is a vital area for improvement in rotational moulding.

Tests on the performance of a range of polyethylene grades [1,4] have shown that a relationship exists between the maximum temperature achieved inside the part and the impact performance of the material. The effect of heating in the presence of oxygen is to form a degraded layer at the inner surface. Additives can offset the effects of oxidative degradation to a degree, but once the layer of degraded material reaches a critical thickness, the strength properties of the part are drastically affected. The brittle degraded layer, which can be seen as a shiny crazed surface, acts as a crack initiation zone. Figure 14 shows a micrograph taken of the cross-section of a degraded part. A layer of material with a different crystal structure can be clearly seen at the inner surface on the right. The onset of this is accompanied by a strong smell and it is this which many moulders use as a guide to the condition of the material after processing. However, measurements of the temperature at which this occurs have shown remarkable consistency regardless of oven temperature, mould material or part thickness. For parts up to 10 mm, at approximately 200°C there is sufficient degradation to cause a sharp drop in impact strength. Beyond this thickness the peak control temperature will be lower due to the extended times for which inner layers are exposed to high temperatures.

By controlling the maximum temperature achieved inside the part, the moulder can consistently produce parts with high strength properties without risk of excessive degradation. Three typical properties for monitoring the quality of parts are considered here:

3.1 Impact Strength

Impact strength is particularly sensitive to overcure of the material during processing. Rotationally moulded parts are essentially thin walled vessels; during impact the inner surface is placed under tension - any defect or weakness in this zone drastically affects the part's strength. Figure 15 shows cold temperature impact results

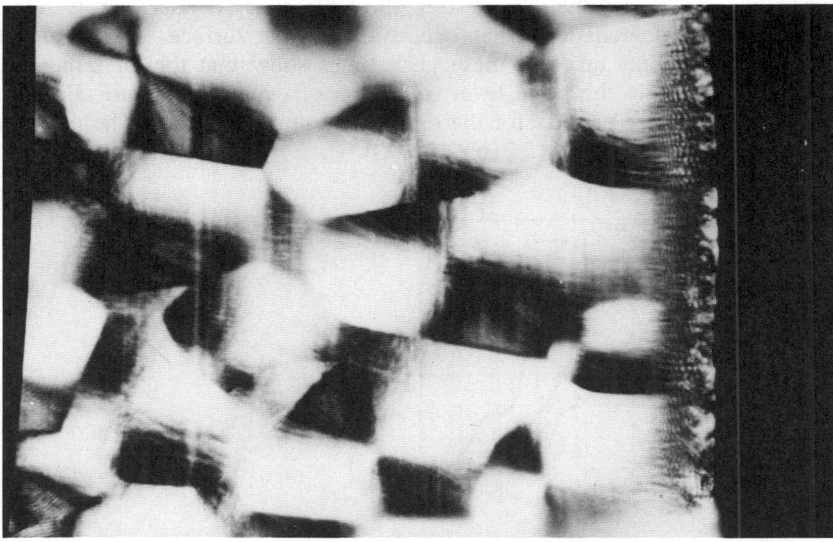

FIGURE 14 Crystalline structure of the cross-section of a 3mm MDPE part showing degraded inner surface

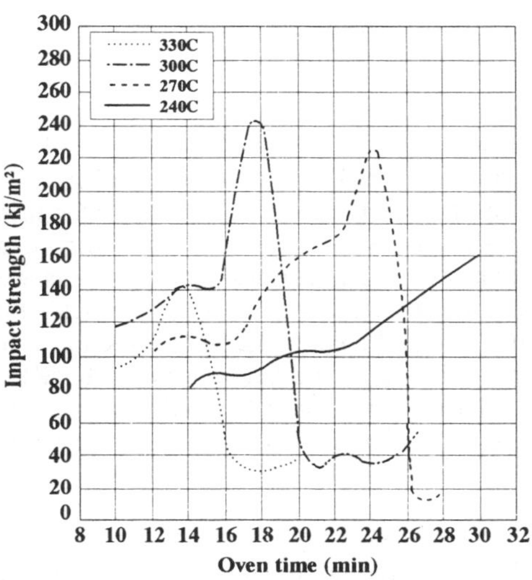

FIGURE 15 Effect of oven time on impact strength at several oven temperatures (MDPE 3mm)

for 3mm thick polyethylene samples which had been produced at a range of oven times for several oven temperatures. For each oven temperature there is a clearly defined time after which the impact strength of the material falls sharply. This is attributable to the degradation which occurs at the inner surface. Internal air temperature measurements taken at these peak times show that the air temperature reaches approximately the same level regardless of oven temperature. Figure 16 shows the temperature profiles for the three maximum conditions. The peak temperatures measured were 222°C, 216°C and 210°C.

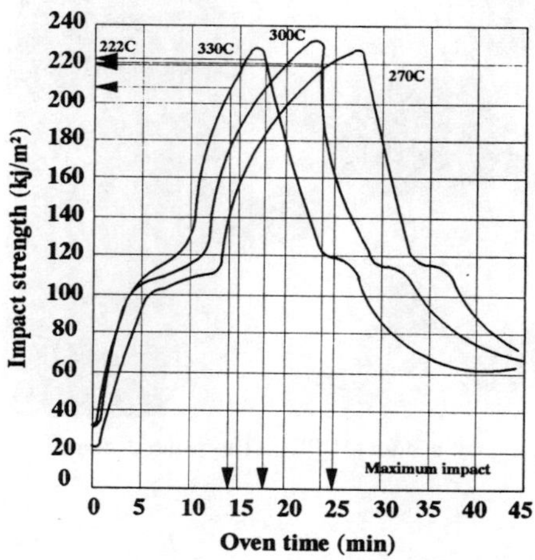

FIGURE 16 Internal air temperature profiles with points of maximum impact shown

3.2 MFI (Surface and Within Wall)

As the plastic material is heated, the flow properties change due to thermal breakdown and oxidative degradation. Melt Flow (MFI) measurements taken at the inner surface of a part also exhibit a sharp downward change due to overcuring. Figure 17 shows the way in which the MFI at the outer surface, inner surface and combined cross-section change with oven time. The surface measurements were made by shaving off a thin layer of material from either face of the part whilst the cross-sectional tests used material right through the part. The MFI at the inner surface falls away to zero very rapidly after oxidative degradation occurs. The MFI at the outer surface falls slowly as the material suffers only thermal degradation in the absence of oxygen. The MFI for the cross-section also falls sharply but not so dramatically as the inner surface.

3.3 Bubble Count

Bubbles are a common problem for rotational moulding as the process naturally entraps air pockets between melting powder particles. These do not move or

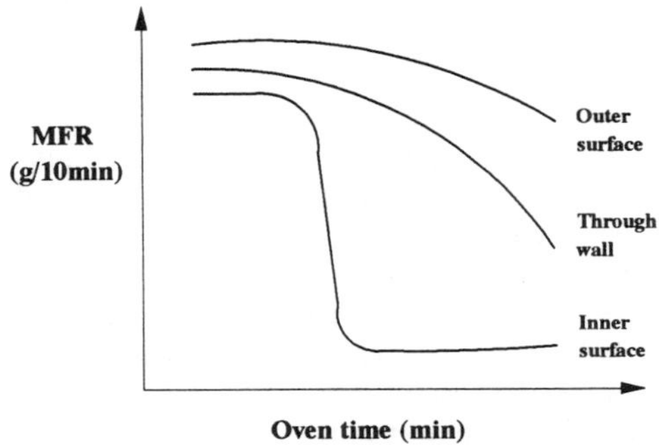

FIGURE 17 Effect of oven time on MFR at inner surface, through complete part and outer surface

migrate within the material but remain static where they are formed and dissolve/diffuse into the surrounding material during heating [5]. However, if the MFI is low then the bubbles will quite often remain even beyond the point at which degradation and detrimental damage has occurred. If the MFI is high then the bubbles will quite easily dissolve/diffuse and disappear completely before the maximum properties are achieved. Figure 18 shows the way in which the bubble count changes with temperature and time. Higher temperatures promote faster dissolution rates.

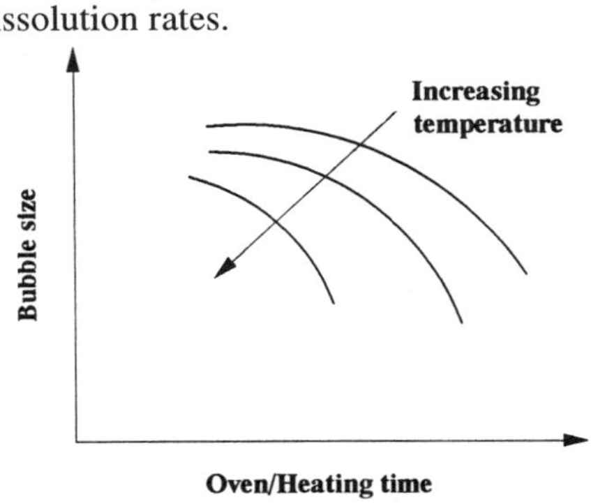

FIGURE 18 Effect of temperature and time on bubble size in rotationally moulded parts

Bubble count has in the past been used as a guide for consistent processing. However, it can be quite difficult to determine the levels of bubbles in coloured parts without destroying the article, and levels may change from position to position in a complex moulding.

4. PROCESSING ADVANTAGES

Real time data from within the mould not only allows the moulder to optimise material properties but also permits greater experimentation with new techniques. Several major areas for immediate benefit are in the balancing of moulds for improved cycle times; multi-layer (multiple shot) processes; control of blowing and crosslinking agents; liquid systems.

4.1 Mould Balancing

Moulders with many different products will often run several different moulds on the same arm of a machine. Where the part thicknesses and moulds are very similar the cycle times will be similar. However, it will be more likely that the moulds will be of different sizes, that the parts are of differing thicknesses or that the actual materials may be different. This can result in problems of different levels of cure; some parts may be properly cured at the expense of overcuring another. Using the processing curves for a range of parts under consistent moulding conditions allows the moulder to establish the best group of moulds to use together to gain the most even curing and most efficient oven use. This may allow, for example, the use of thin parts in thick aluminium moulds alongside thicker parts in thin steel moulds, or polypropylene parts alongside polyethylene parts.

Figure 19 shows a trial which examined four different moulds simultaneously. All four were apparently similar in wall thickness and cycle time. However, during processing it immediately became obvious whilst using Rotolog that one of the moulds (line 4) was lagging behind the other parts. The cycle had been set to allow this part to be properly cured which meant that the other three parts were overcured, reaching internal temperatures in excess of 220°C. To balance the arm properly would have required a fourth mould with a similar performance to the first three moulds. However, as a temporary measure the shotweight in the fourth part

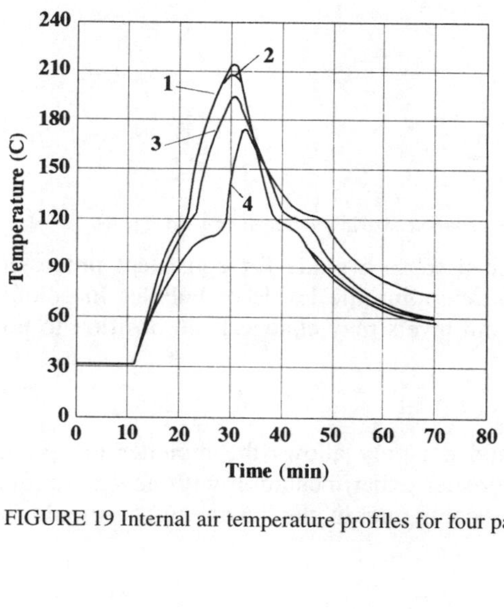

FIGURE 19 Internal air temperature profiles for four parts moulded simultaneously – original cycle

was reduced to bring the wall thickness into line with the other parts. The heating cycle was then reduced and the cooling cycle altered by adding a water spray. Figure 20 shows the revised cycle. The three parts were still slightly overcured, but to a lesser degree (207°C), and the fourth part was still properly cured. The point at which demoulding occurred was also improved so that the overall improvement of cycle time was 8 minutes (14%) with potentially greater savings if a fourth 'balanced' part could be found.

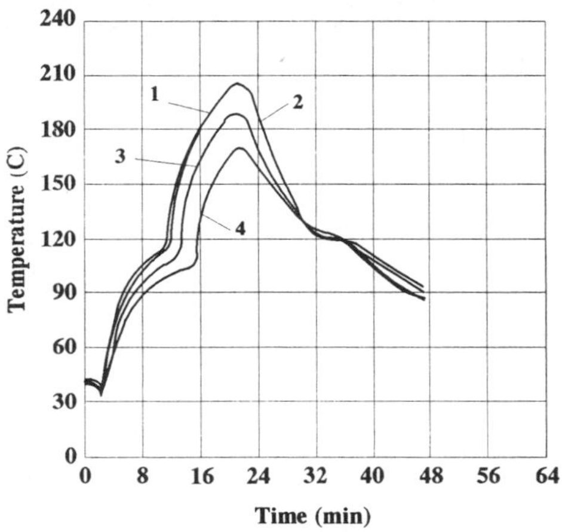

FIGURE 20 Internal air temperature profiles for four parts moulded simultaneously – revised cycle

4.2 Additives

Control of the internal air temperature within the mould is vital where additives are used. Even standard grades of materials have a cocktail of heat and UV stabilisers which must be preserved. By ensuring that the material is not overcured, the moulder is ensuring that the levels of these compounds remain sufficiently high for the expected lifetime of the part. Other additives such as blowing and crosslinking agents must be deliberately degraded. These will typically require activation temperatures in the range of 180-200°C. To ensure that these compounds have been fully used and to achieve consistent results, the moulder requires a knowledge of the temperature reached.

Figure 21 shows the effect of pre-blending blowing agent with a standard polyethylene material. Above a certain temperature, the blowing agent breaks down to release gases which produce a foamed part. This requires good dispersion of the agent and an activation temperature above the melting point of the polymer. During processing, the material adheres in the normal fashion trapping the dispersed blowing agent through the wall thickness. Once the activation temperature (160°C +) for the blowing agent is reached the chemical breaks down and an exotherm occurs as the part foams. During cooling, this exotherm and the enhanced insulation properties of the part retain heat inside the part and reduce the rate of cooling quite considerably. Figure 21 shows how the internal temperature remains high and falls more slowly at the onset of cooling.

Figure 22 shows the internal air temperature profile for a crosslinking grade of material. Crosslinking agents improve the strength and stiffness of standard polyethylenes and produce materials with improved permeation resistance. During processing, the crosslinking process does not occur until around 180°C (depending on the grade used) and produces no major exotherm during reaction. Thus the internal air temperature curve exhibits no major difference from a standard grade. The information required by the moulder is to ensure that the activation temperature has been exceeded consistently for all mouldings.

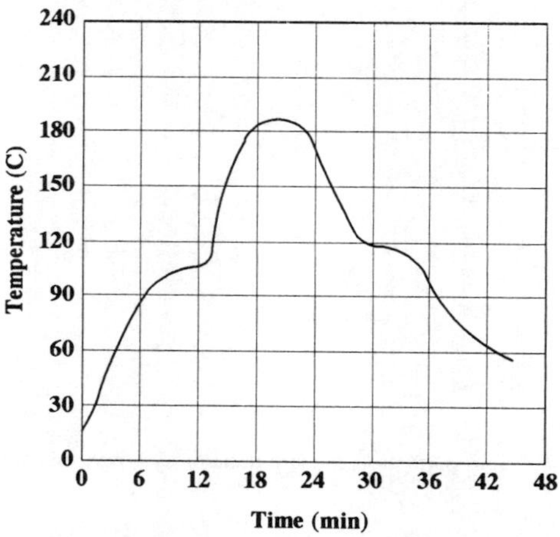

FIGURE 21 Internal air temperature profile for MDPE premixed with blowing agent

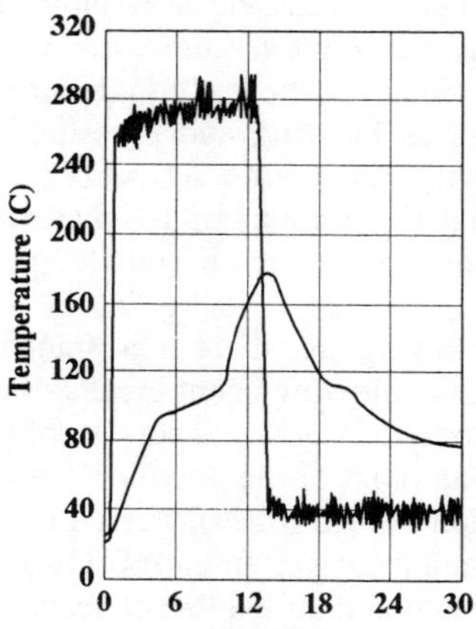

FIGURE 22 Oven and internal air temperature profiles for crosslinkable Stamylex (XLPE) (3mm)

4.3 Multi-layer Technology

Multiple layers in a moulding offer many potential advantages such as: increased stiffness where a solid skin and foam are combined; improved barrier properties and permeation resistance by using a thin inner or outer layer of low permeability material; cost savings where an expensive material can be laid down on a substrate of cheaper standard material. Developments in multilayer parts such as these have been hampered not only by the physical difficulties of adding a second shot of material but also by not knowing precisely when to add it. This can now be determined easily from observations of the internal temperature of the mould, so removing much of the trial and error previously involved.

Figure 23 shows a two stage process where a skin of standard polyethylene material is combined with an inner layer of another material. The first shot of material moulds in the normal fashion with a plateau indicating the points at which the material first adheres to the mould and when it has finished adhering. At point ‘P’, when this layer has cured, the mould is removed from the oven and the second shot of material added. This has the effect of cooling the inside of the part as shown at

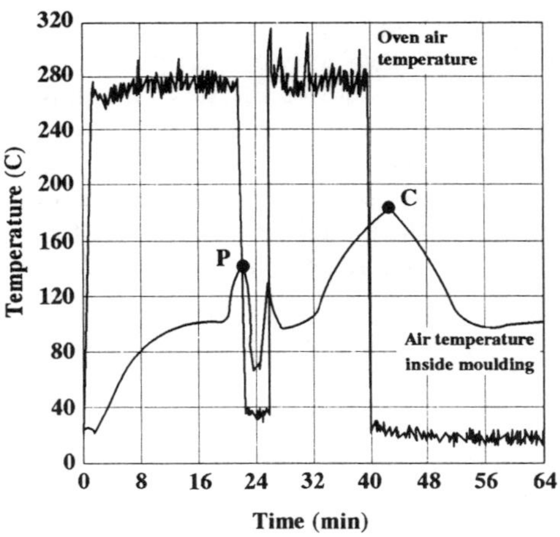

FIGURE 23 Oven and internal air temperature profile during a two shot moulding process

point ‘Q’. When heating is resumed, the second layer fuses in the same manner as the first and the moulder is able to control the internal temperature to ensure optimum properties in the second layer (point ‘C’). When producing very thick parts the moulder must be careful not to thermally degrade the outer layers at the expense of optimising the properties of the inner layer.

4.4 Liquid Systems

There are obvious differences between the behaviour of liquid and powdered materials. These mainly lie in the manner in which material is distributed and the way in which it builds up on the inner surface of the mould. Liquids, particularly viscous ones like PVC plastisols, coat the mould immediately and, if the part to be made is sufficiently thin, all the material can be distributed in the first few rotations. For thicker parts and less viscous materials, the plastic lies in a pool at the bottom of the mould and flows over the surface in a manner similar to powder. Build-up on the mould surface occurs as the viscosity of the material decreases due to chemical reactions. Tests using liquid PVC have shown that for thick parts (>5mm) there is a plateau exhibited around 65°C during build-up (see Figure 8). This is less important, however, than the maximum temperature achieved which is related to the level of cure of the material. At around 170°C the material 'gels' to form a well cured part.

Other materials have been tested, for example Nyrim. This material cures quite rapidly once the temperature and reaction conditions have been achieved so that it is critical that the material is distributed before this occurs. The processing reaction can be seen quite clearly as an exotherm [6].

Moulding of some liquid materials requires that two shots are used to ensure even distribution. This results in quite a complex process with several more variables added to an already complicated quality control problem. Fig 24 shows temperature measurements taken during the moulding of a nylon part produced using two shots. The upper curve shows the oven temperature whilst the lower shows the temperature at the inner surface of the mould. As this was the first cycle for the day, the mould was cool at around 25-30°C. As a result a longer first oven cycle was required to bring the mould to the target temperature of 150°C for the addition of the first shot of material. The temperature of the mould stabilises as the first shot of material is added and then rises as an exothermic reaction takes place. This reac-

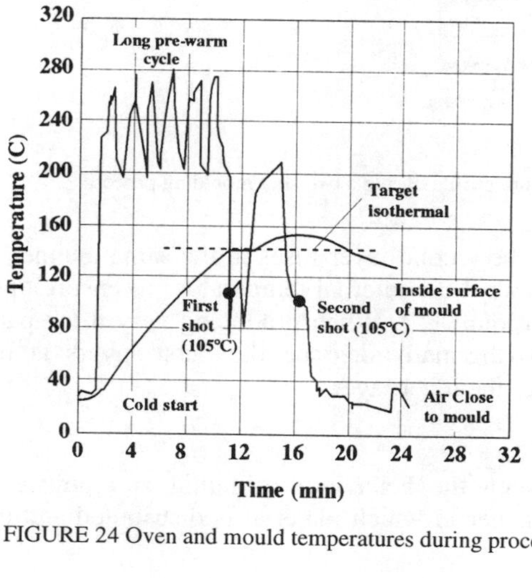

FIGURE 24 Oven and mould temperatures during processing of a liquid Nylon part

tion takes place as the mould is moved to a second oven stage. Initially the second oven temperature had been set to 190°C. After a short second heating period, the second shot of material was added and this was distributed and allowed to react as the mould rotated in the cooling bay. The reaction inside the mould should ideally take place under isothermal conditions.

Figure 25 shows a later test for the same part. The first oven cycle is considerably shorter as the mould is now hot before entering the first oven. The second oven temperature has been reduced to 170°C to reduce the heat input for the isothermal reaction. Note also that, after assessing 5-6 parts, the target temperature for the addition of the first shot of material was reduced to 140°C.

5. MACHINE IMPROVEMENTS

Machine design has been improving rapidly over the last few years with major reductions in cycle time by increasing oven and cooling bay efficiencies. These improvements, however, have been made as machine improvements without feedback of their effects on the mould surface or the plastic inside the mould. Real time data from the mould, both inside and outside, can provide vital data on the effects of various changes. Data on how the temperature varies across an oven, on how the temperature recovers after the oven doors have been opened, on how the cooling bay and ambient environment temperatures are related and on how the cooling bay temperature is affected by various cooling media can provide valuable guidelines for machine design.

Probes can be placed close to the outer surface of the mould to examine the environment that the mould experiences as it moves about the oven and cooling

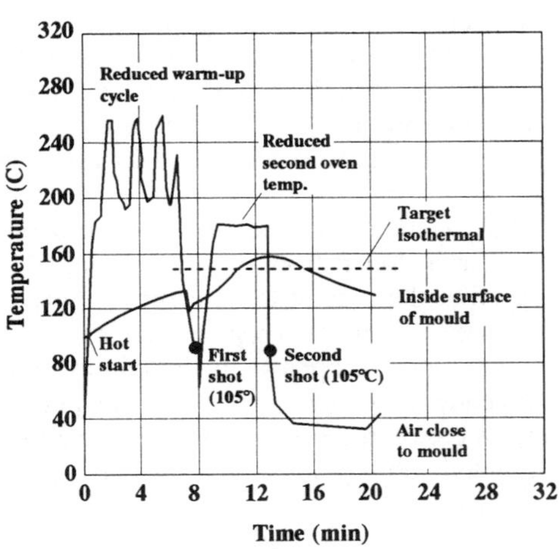

FIGURE 25 Oven and mould temperatures during processing of a liquid Nylon part

areas. This can show how the temperature fluctuates across an oven as each side of the mould passes in turn through the hot air inlet to cooler areas of the oven. Depending on the size of the oven and how much free space the mould takes up within this, the fluctuations can be as much as +/- 60°C around the oven set point (see Figure 26). Improved air flow could produce a more homogeneous oven environment.

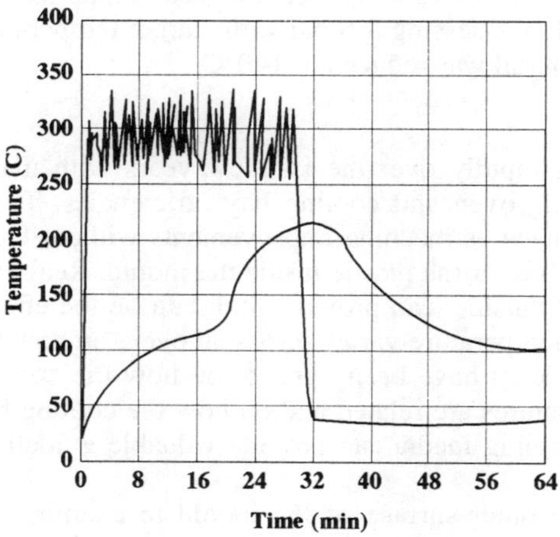

FIGURE 26 Oven temperature profile for a machine without a modulating burner (cycle shown for 10mm part)

Some ovens use an overshoot of temperature to speed up recovery of the oven temperature after the doors are opened to allow a cold mould to enter the oven. Figure 27 shows just such a cycle on a large moulding machine. This can improve cycle times but the peak temperature achieved must be limited to ensure mould and machine component lifetimes are not detrimentally reduced. Fig 27 also shows the effect of using a modulating burner.

During cooling it can be shown that, when a mould enters a closed cooling bay, the ambient temperature will rise as heat is drawn from the mould. Figs 27 and 28 show a mould which has been delayed at the rear of the machine before entering the cooler. As the mould enters the cooling bay there is a sharp increase in temperature of about 20°C. If the cooler doors are left open then the cooling air is that much cooler and the cooling efficiency is higher. Also, on independent arm carousels, delays at the holding station between the oven and cooler can be eliminated by the use of an externally mounted cooling fan so that the cooling cycle can begin immediately the mould leaves the oven.

Adding a spray of cooling water has the effect of reducing the ambient cooling temperature inside the cooling bay and also of increasing the rate of heat transfer

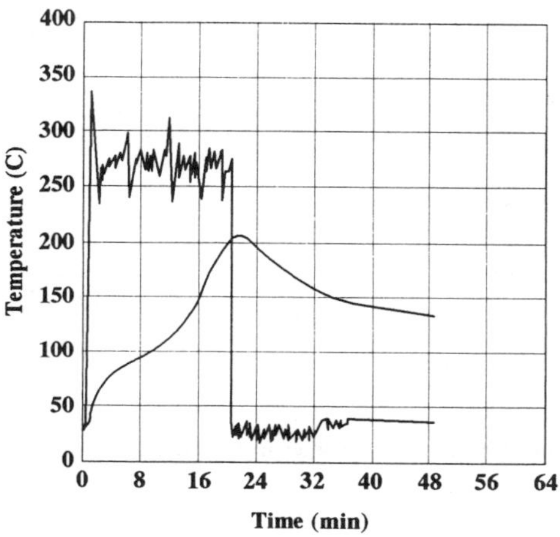

FIGURE 27 Oven temperature profile for a machine with a modulating burner (cycle shown for 6mm part)

from the mould. Figure 28 shows how a water mist after 69 minutes causes a sharp drop in environment temperature.

6. QUALITY CONTROL

Quality control and process monitoring will be a major area for expansion in rotational moulding. With the advent of computer workstations to control machine movements, the capabillty for data collection and storage is available. Information such as processing conditions and mould material data are already stored automatically on some modern machines. However, information on the state of cure of the part, mould temperature or machine oven performance are not. Data such as this is vital for true quality control and performance analysis.

Information such as this can be used for statistical analysis. Continuous output from each mould would provide complete diagnostics on part quality and machine efficiency during production. As each cycle is completed so the information would be stored in a central fileserver. At the end of a shift (or continuously during shifts if required) the data could be compiled to produce a breakdown of production and performance. This could be linked to any point in the factory for remote monitoring. Trends and problems could be more quickly identified and changes in conditions more easily controlled.

With more demands being made by customers for traceability and conformity, the moulder could now provide a processing trace for each part to show that it has

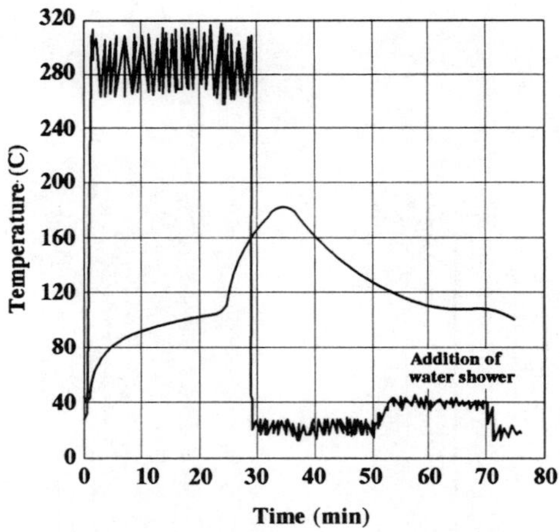

FIGURE 28 Oven and internal air temperature profiles for part cooled under ambient conditions followed by water

been cured to a standard level. This type of documentation would be a major boost to the general image of rotational moulding, particularly for moulders working to meet technically demanding markets.

It is also evident that the next stage of process control in rotational moulding will be direct machine control from the temperature inside the mould. This has already been shown to be possible using, for example, radio frequency transmission of data from the mould to the machine controller. Thus moulds will automatically leave the oven at precise pre-set points independent of transient variables such as ambient temperatures, delays in closing oven doors, incorrect charge of material in the mould, variations in quality of the plastic material, etc. Rotational moulding will thus have moved into the high technology process control previously only associated with injection moulding and extrusion blow moulding.

REFERENCES

[1] Crawford, R.J., Nugent, P.J. and Wang, Xin, 'Prediction of optimum process conditions for rotomoulded products.' Int. Polym. Proc., VI (1991) pp. 56-60.

[2] Nugent, PJ., 'A study of heat transfer and process control in the rotational moulding process for polymer powders.' Ph.D Thesis, Queen's University, Belfast, Sept. 1990.

[3] Nugent, P.J., and Crawford, R.J., 'A new process control system for rotational moulding', Plast., Rubbers and Comp., Proc. and Appl., 17, 1 (1992) pp 23-31.

[4] Nugent, P.J., and Crawford, R.J., 'Impact strength of rotomoulded PE products', Plast., Rubbers and Comp., Proc. and Appl., 17, 1 (1992) pp 33-41.

[5] Crawford, R.J., 'The formation and removal of bubbles in rotationally moulded PE', Plast. and Rubber: Proc. and Appl. 7(1987) pp 85-99.

[6] Harkin-Jones, E., 'Rotational Moulding of Liquid Plastics', Ph.D thesis, Queen's University, Belfast (1992).

CHAPTER 10

Pin-holes and Bubbles in Rotationally Moulded Products

A.G. Spence and R.J. Crawford

1. INTRODUCTION

As the awareness of rotational moulding and opportunities for the industry continue to develop, there is an ever increasing demand for improved product quality. Mouldings with a poor surface finish sometimes reflect the quality of mould used. However, the main concern about quality is usually related to the occurrence of surface pores and bubbles within the walls of rotationally moulded articles. The formation of bubbles and surface pores is an inherent problem with the rotational moulding process, and has plagued the industry for the past 30 years. Bubbles in mouldings are generally undesirable since they reduce stiffness and impair the appearance of a product. They can also cause hygiene problems in some products if they occur at the surface.

This chapter reviews the processes involved in the formation of bubbles in rotationally moulded products[1]. The relative effects of the various factors will be assessed and methods are identified whereby the problems can be reduced or removed completely.

2. BUBBLE FORMATION

While the presence of bubbles in rotationally moulded products has been referred to by several authors, only a few researchers have made a serious attempt at explaining how bubbles are formed and how they are removed. In an initial attempt at this problem, Rao and Throne[2] drew extensively from the theories of metal sintering and glass densification. It was suggested that the formation of a homogeneous melt from powder particles involves two distinct steps:

1. The particles stick or fuse together at their points of contact. This fusion zone grows until the mass becomes a porous three-dimensional network with relatively little density change. This is referred to as **sintering**.
2. At some point in the fusion process, the network begins to collapse into the void spaces. These spaces are filled with molten polymer that is drawn into the region by capillary forces. This is referred to as **densification**.

Rao and Throne used this model to explain features such as surface porosity of mouldings. They suggested that these occur because the voids in the powder interior that are formed upon densification are pushed ahead of the melt front to the free surface. However, the buoyancy, capillary and hydrodynamic forces, which would cause the interior void to move away from the mould surface, are not strong enough to overcome the surface tension force required to pull the surface void away from the mould surface and into a bubble. It was concluded, therefore, that surface porosity is intrinsic to the process.

Ten years later, Progelhof et al[3] further developed the powder densification theory. Their updated theory suggested that as the powder is heated, the particles become sticky and adhere to each other, and upon further heating, the particles fuse together or densify to form a unitised structure. From observing hot plate experiments, Progelhof and his associates concluded that as the heating process continues, the solid/melt interface moves upwards, the top of the free surface of the powder drops, and with time the powder melts completely. Progelhof also observed the formation of voids. These appeared to be a result of a bound inclusion of the space between individual particles, and the most striking point was the slow movement of voids to the free surface. There appeared to be some coalescence of voids, but with time the voids diminished in size.

Powder densification was also considered by Kelly[4] of Du Pont Laboratories, Canada. Kelly suggested that air bubbles are trapped in the polymer during melting, and decrease in diameter as the melt temperature increases. The high viscosity of the melt prevents movement of the bubbles. At a high enough melt temperature, the air in the bubbles begins to dissolve into the polymer. Oxygen has about twice the solubility of nitrogen in polyethylene. At high temperatures, the oxygen is further depleted by direct oxidation reactions with polyethylene. The depletion of oxygen reduces the bubble diameter. The laws of surface tension dictate that the pressure inside the bubble has to increase as the diameter decreases. The increase in pressure causes the nitrogen to dissolve in the polymer, thus the bubble diameter is further reduced and this chain of events repeats until the bubble disappears. As cycle time increases, the size and quantity of bubbles decrease. At very long cycle times, the moulded part will have no bubbles. However, in such cases the impact strength will be low because the polyethylene has oxidised.

Crawford and Scott[5] carried out hot plate tests on powders, using video equipment to record and examine the processes of melting in detail. They based the conditions on earlier measurements made during actual moulding trials. Vacuum dried powders confirmed that solvent residue or moisture were not involved in the formation of bubbles. The video recording allowed the formation and subsequent dissolution of bubbles to be recorded and modelled. They described the melting and collapse of the powder as heat is applied from the base in terms of two bulk movements. Firstly, the progression of the melt front through the powder mass to the free surface and the collapsing of powder particles into the melt. It is the latter that tends to trap air. A force analysis on a typical bubble shows that the apparent viscosity of molten polyethylene is so large that the buoyancy forces acting on it are insignificant. They also stated that the trapped air will diffuse into the surrounding polymer mass, and produced measurements to support this. The initial size of the bubble has a significant effect on the rate at which it dissolves, as the surface area-to-volume ratio is inversely proportional to the diameter. They found that this initial size was dependent on the size of the particles, and that the number of bubbles was dependent on the particle size distribution and the MFI of the material. Crawford and Xu[6] complemented Scott's earlier work on bubble analysis. They derived a semi-empirical relationship between bubble diameter ratio, temperature and time. From this relationship the bubble diameter was predicted as a function of time and melt temperature.

Therefore, it can be concluded that bubbles form due to the encapsulation of air pockets between powder particles, as they melt and fuse together (see Figure 1). As the melting process continues, these bubbles remain stationary, due to the high viscosity of the molten polymer. The bubbles then slowly diminish in size and may disappear depending on the viscosity of the material, the heating cycle time and the melt temperature. The initial bubble formation stage is critical in that the size of bubbles which form can be influenced by the powder's characteristics and by the surface of the mould.

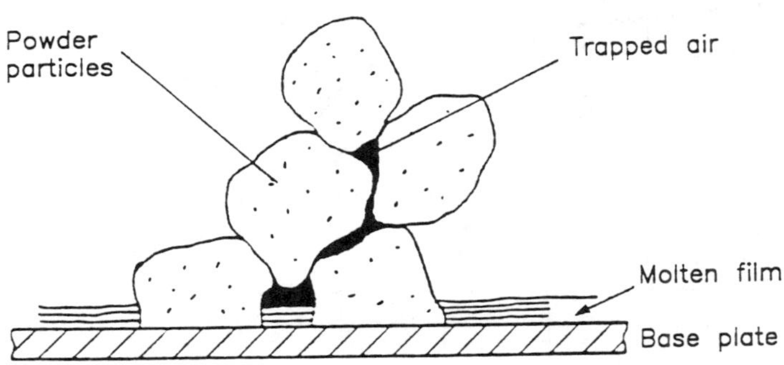

FIGURE 1 Bubble formation in a melting powder

3. THE EFFECT OF THE MOULDING MATERIAL ON BUBBLES

One of the most influencing factors when considering bubbles in rotomoulded products is the type of polymer being used. A common feeling in the industry is that some materials are more prone to producing parts with bubbles than others. How can this be, as bubbles must form in all materials due to pockets of air being trapped as the powder particles melt and fuse together? The answer lies with the rheological properties of the polymer. Bubbles do form in all materials during the melting process; however, some materials allow the air molecules contained inside each bubble to diffuse out of the polymer more easily than others. This phenomenon was observed using a hot plate apparatus by comparing materials with different MFI's[1]. The hot plate test provides a means by which the formation and removal of bubbles can be observed in various powdered materials while they are being heated on a metal plate.

3.1 Melt Rheology

As the polymer temperature rises, it changes from solid particles to a melt state, containing small bubbles typically 0.3mm to 0.5mm in diameter. Figure 2 illustrates how the diameters of bubbles change with increasing melt temperature for six different grades of polyethylene with melt indices ranging from 3.2 to 25g/l0mins. It may be seen that all materials contain bubbles during the initial melt-down phase. As the melt temperature increased, bubbles diminish in size until after a period of time, they disappeared completely. It should be noted that the rate at which bubbles disappear increases as the Melt Flow Index of the material increases.

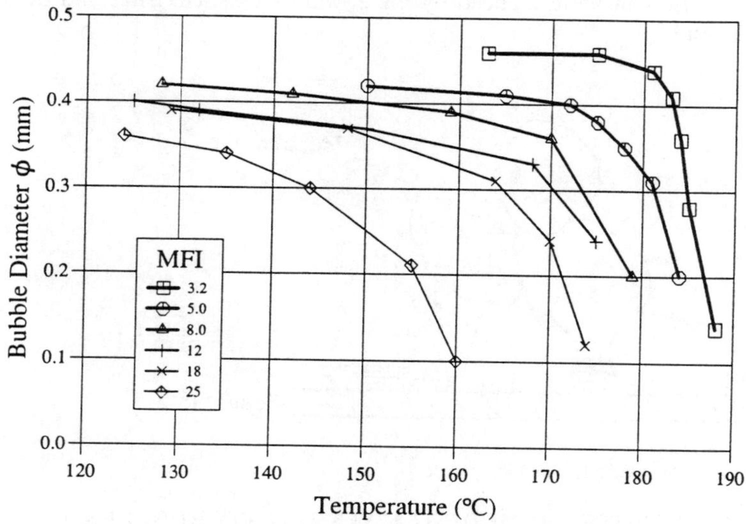

FIGURE 2 Hot plate tests

These observations may be explained by the fact that as the MFI of the material decreases, the molecular chain length increases. Longer chain lengths not only mean that the melt is more viscous, but also, that the passage of air molecules through the melt is more difficult. In practical terms this means that during the rotational moulding cycle there may not be enough time for bubbles contained in a more viscous material (e.g. 3.2 MFI) to diffuse out of the polymer when it is in its molten stage. However, with higher MFI materials (i.e. shorter chains), air molecules have an easier path to the free surface, and so bubbles have sufficient time to diffuse out of the polymer during the rotational moulding cycle.

The general process of the removal of bubbles from each material is similar, and may be influenced by three factors. Firstly, as the melt temperature increases, this causes the pressure inside the bubble to increase. The laws of surface tension dictate[7] that if the bubble pressure increases, then its radius must decrease by an inversely proportional amount. Secondly, as the melt temperature increases, the polymer structure breaks down and so bonding becomes weaker, allowing the air molecules an easier passage through the melt. Thirdly, at high temperatures, the oxygen contained within bubbles is further depleted by direct oxidation reactions with the surrounding polymer. This corresponds to a change in colour of the polymer from natural to a creamy brown.

During the rotational moulding process, the mould is rotated at relatively slow speeds. Unlike other polymer processing techniques this causes low shearing forces between the molten polymer and the mould wall. Low shearing of the polymer causes the viscosity of the polymer to rise and so bubble diffusion becomes more difficult. Moulders should be aware that the material's MFI value represents only one point on the material's viscosity curve at 190°C. This value is not relevant to the rotational moulding process as, during moulding, the temperature of the material is changing continuously and the shearing forces experienced are low - certainly much lower than the values which occur during the Melt Index test.

The viscosity of the polymer can be influenced by four main factors[8]: molecular weight, shear rate, temperature, and pressure. Table 1 shows the average molecular weight for a range of materials with different Melt Flow Indices. From Table 1 it can be seen that as the Melt Flow Index increases, the average molecular weight decreases. This can be explained by the fact that the molecular chain length is decreasing.

Table 1
Variation of Average Molecular Weight with Melt Flow Index

MFI (g/10 mins)	Type	Colour	Molecular Weight
3.2	MDPE	Natural	100,300
5.0	LDPE	Natural	89,550
8.0	MDPE	Natural	80,400
12	MDPE	Natural	70,000
18	MDPE	Natural	64,000
25	MDPE	Natural	59,000

Figure 3 illustrates how polymer viscosity, for a range of natural materials, varies with temperature at a constant shear rate of $0.1s^{-1}$. As expected, higher molecular weight materials correspond to higher viscosity values. However, the viscosity of the polymer varies significantly with temperature, particularly for the higher molecular weight materials. During rotational moulding, it is the more viscous (i.e. higher molecular weight) materials that tend to produce mouldings which contain bubbles and surface pores. The information in Figure 3 correlates well with the data presented in Figure 2 and complements what has been proposed previously. It is suggested that air molecules will begin to diffuse out of a natural material when the viscosity has reached a suitably low level. Therefore, to accelerate the diffusion process of air bubbles from a high molecular weight material, the viscosity must be reduced, either artificially or otherwise.

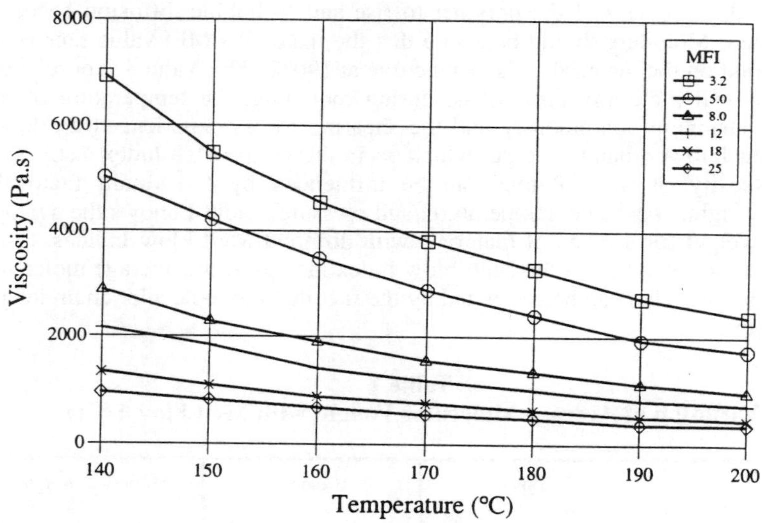

FIGURE 3 Viscosity profiles for natural materials at a $0.1s^{-1}$ shearing rate

3.2 Viscosity Reduction Techniques

One method of reducing the viscosity of a polymer is to increase its temperature. However, this method has been shown to have a detrimental effect on the polymer's structure. During a series of rotational moulding trials[1], the mould was removed from the oven at a range of peak internal air temperatures (125°C to 250°C). It was not until the peak internal air temperature inside the mould reached 250°C that bubbles and surface pores were finally removed from this particular material. However, impact tests (see Figure 4) showed that although bubbles and

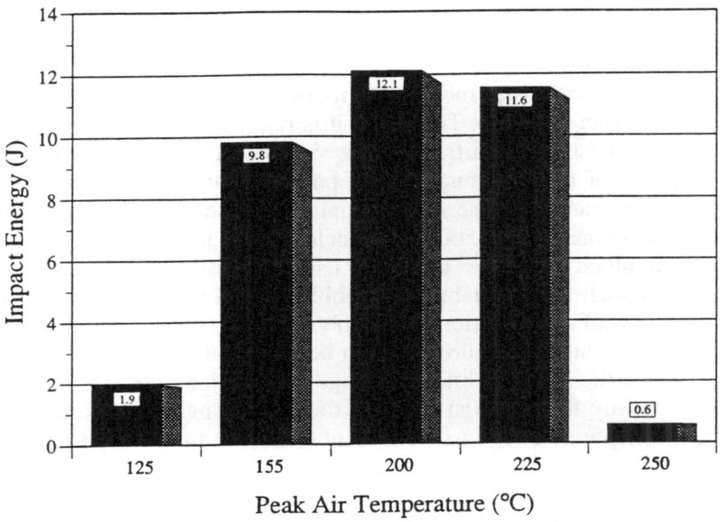

FIGURE 4 Impact energy for a range of peak internal air temperatures

surface pores had been removed, the polymer structure had degraded due to oxidation. Another disadvantage to this method of removing bubbles is the increase in cycle time experienced when higher peak internal air temperatures are used.

Another method to reduce the viscosity of a polymer is to increase the rate of shear. Unfortunately this method for reducing the viscosity of the polymer is not suitable for the rotational moulding process. Rotational moulding is restricted to low shearing rates because the mould must rotate slowly, to minimise the inertia forces generated as the relatively heavy, irregularly shaped mould and its contents are rotated.

An alternative method for changing the rheological properties of the polymer is by introducing an additive. Small quantities (1 or 2%) of materials such as waxes or stearates are known to reduce the viscosity of a polymer melt. These additives tend to have a low viscosity in their melt state, which helps to lubricate the flow of the more viscous polymer. One of the most important criteria when selecting a suitable wax or stearate is the temperature at which it melts. If the selected additive melts at a similar temperature to the polymer, then the structure of the polymer may be severely weakened due to the inclusion of the additive particles. However, if the additive melts before the polymer, then it tends to wet the inner surface of the mould and form a pool of low viscosity liquid. When the polymer begins to melt, the additive pool tends to flow into the surface cavities between melting powder particles, and so reduces, or eliminates entirely, surface porosity in the moulded part. For coloured parts this is all that is necessary to produce what appears to be a bubble free product. Additives which have a higher melting temperature than the polymer tend to be not so effective.

3.3 Powder Characteristics

Bubbles which form in rotationally moulded parts occur as a result of air trapped between individual powder particles. Therefore, it is reasonable to suggest that the powder's characteristics (such as particle shape, size and distribution) will affect the formation and size of bubbles, and surface pores in particular. The effect of particle size was investigated[1] by series of mouldings with varying percentages of fine (> 300μm) and coarse (300-500μm) particles. After moulding, the surface porosity of each moulded part was observed using an image analyser, which provided information such as the quantity of bubbles, the size of bubbles and the percentage area of the wall cross-section with bubbles.

From the results illustrated in Figure 5, it can be seen that as the percentage of fine particles increases, then the maximum, average and median pore size decrease. This can be explained by the fact that finer particles trap smaller pockets of air, which result in smaller pores. The most significant change in the average pore size and the median pore size occurs when the percentage of fine particles is between 0 and 10%. This is due to the sieving action which takes place during rotational moulding. As the powder tumbles inside the mould, the fine particles tend to be

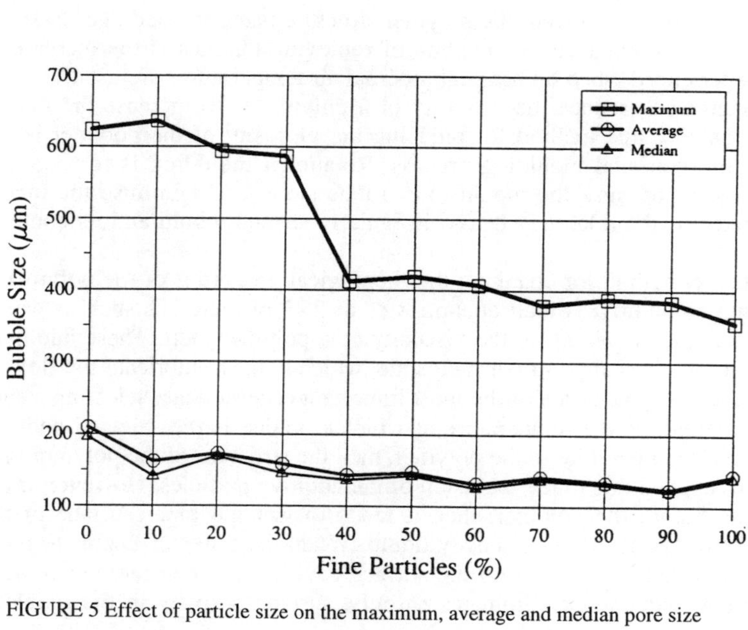

FIGURE 5 Effect of particle size on the maximum, average and median pore size

sieved to the inner surface of the mould. The finer particles therefore melt first against the mould. If this is the case, then the inclusion of more fine particles will have less of an effect on the size of the surface pores, but increasingly influence the size of bubbles contained in the wall of the part. In general, smaller particles produce a greater number of bubbles which are smaller in size and cover a smaller percentage area of the product. Smaller bubbles are more desirable as they contain a smaller volume of air, and therefore require a shorter period of time to diffuse out of the melt.

4. MOULD CONSIDERATIONS

The quality of any rotationally moulded item is governed principally by the quality of the mould used. The mould material and, more particularly, the quality of its internal surface finish are two of the most important factors which influence the appearance of a product. Typically, moulds are made from mild steel, stainless steel, cast aluminium or electroformed nickel/copper. The choice of mould material depends on the particular application, the number of moulds required and the surface finish necessary. Another important factor is the type of release agent used.

4.1 Mould Surface Finish

The effect of the surface finish of the mould on the surface porosity of the moulded product has been investigated[1] using aluminium and mild steel. These two materials were polished to various degrees from their virgin state and then heated and used to melt polymer powder. Examination of the samples produced showed that the degree of mould surface finish had little or no effect on the quantity of surface pores on the moulded product. In particular, these tests highlighted that there were no advantages in producing a highly polished internal surface on a mould, in terms of the surface porosity of the product. However, there is an improvement in the general surface texture of the moulded product.

4.2 Mould Material

The effect of six different mould materials on the surface porosity of the moulded product was also investigated[1]. The materials used included mild steel, stainless steel (mirror finish), aluminium, mild steel chrome plated, brass (mirror finish) and copper. The hot plate was used to heat the metal plate on which the polymer powder was melted. During the hot plate tests, the effect of each material was investigated both with and without the application of a release agent. From the tests it was found that the surface porosities experienced for each material with a release agent were quite similar, showing a high density of pores. However, when no release agent was present the surface porosity experienced for each material was significantly reduced. Figures 6 and 7 illustrate how the surface porosity produced by mild steel was influenced by the application of a release agent. The difference in surface porosity experienced can be explained by the surface tension effects.

The probability of the molten polymer wetting the surface of a metal mould depends on the relative surface energies of each of the combining surfaces. If the

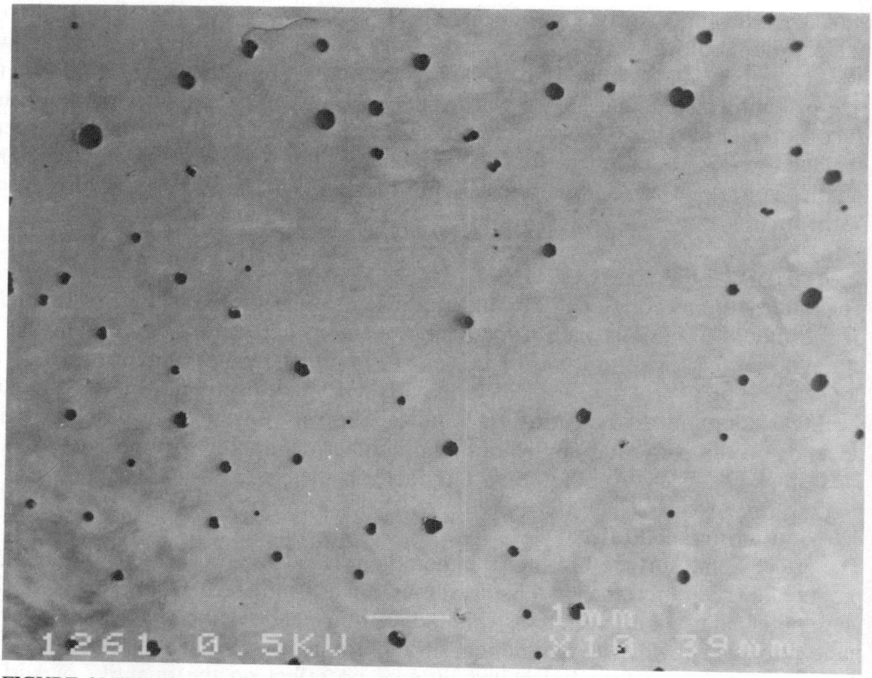

FIGURE 6 Mild steel without release agent (magnification x 10)

surface energy of the mould is greater than the surface energy of the molten polymer, then the polymer will wet, and be attracted to the mould. If the surface energy of the mould is less than that of the liquid polymer then the opposite is true. Values of surface tension for polyolefins lie in the region of 20-50 dynes/cm[9], whereas for metals, the values are much greater[10] (typically 100's dynes/cm). Therefore, the polymer should, in theory, wet and spread evenly over the surface of the mould and result in few surface pores. When a release agent is present, the surface tension of the metal is significantly reduced and thus more surface pores would occur. This is borne out by Figures 6 and 7.

In practice the bare metal (i.e. without release agent) is not as good at reducing surface pores as would be predicted by theory because an oxidation layer forms on the surface of the metal, which reduces its apparent surface energy/tension. During the hot plate test the oxidation process is greatly accelerated due to the elevated temperature involved[11].Measurement of the surface tension of each metal section showed that oxidation had taken place. Although the presence of a release agent increases the amount of surfaces pores, clearly it is necessary to facilitate easy release from the mould. These results, however, highlight the importance of choosing the correct release agent and ensuring uniformity and consistency in its application.

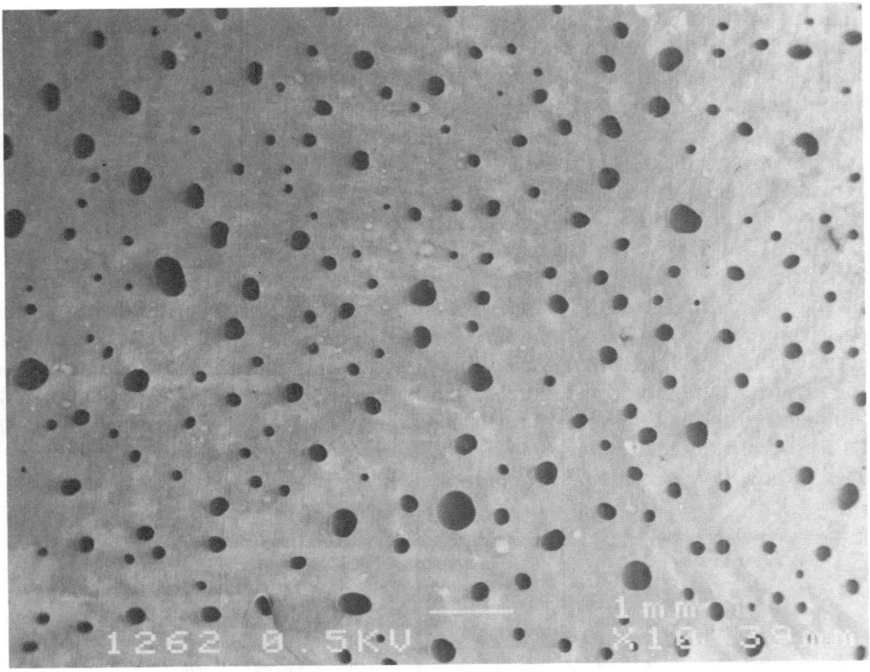

FIGURE 7 Mild steel with release agent (magnification x10)

There are many different types of mould release agent used in the rotational moulding industry. Tests[1] using 6 common release agents showed that the surface porosity experienced for each release agent was quite similar. It was found that the similar effect of each release agent correlated to the surface tension measured for each mild steel plate coated with release agent. It would appear that no "ideal" release agent exists which can provide acceptable non-stick characteristics and also reduce the surface porosity of the moulded product. By their very nature, release agents invariably increase surface porosity, as they reduce the force which attracts the polymer to the mould.

4.3 Mould "Hot-spots"

Rotational moulders have observed that some moulds produce parts which contain areas with no surface pores, while other areas of the same moulding have surface pores. The industry attributes this observation to "hot-spots" or areas of the mould which conduct more heat. Experiments using a thermal imaging camera[1] have highlighted that some moulds can have areas which become locally hotter than other areas of the same mould. This is generally due to variations in the thickness of the mould wall. A thinner area on the mould will allow heat to pass through to polymer more quickly. This will cause a greater build-up of polymer material adjacent to the thinner area of the mould. The mould thickness also governs the temperature of the polymer during the moulding cycle. As the mould thickness decreases, the polymer is subjected to higher temperatures for longer periods of time. This gives the bubbles and pores more time to diffuse through a less viscous melt. Therefore, variations in mould thickness, or variations in the thermal conductivity of specific areas of the mould, will lead to hot or cold spots and this is likely to result in variations in the surface porosity of the moulded part.

5. BUBBLE REMOVAL USING PRESSURE

Pressurising the inner atmosphere of the mould provides the rotomoulder with an alternative method to remove bubbles and surface pores. By introducing a small positive pressure into the mould, after the polymer has melted, bubbles and surface pores can be removed. Once the bubbles have formed in the melt, the introduction of pressure to the mould will act as a compressive force causing the gas molecules contained in the bubble to diffuse through, and out of the polymer (see Figure 8). This process may take only a few seconds to remove all bubbles, depending on the level of pressure applied and the characteristics of the material.

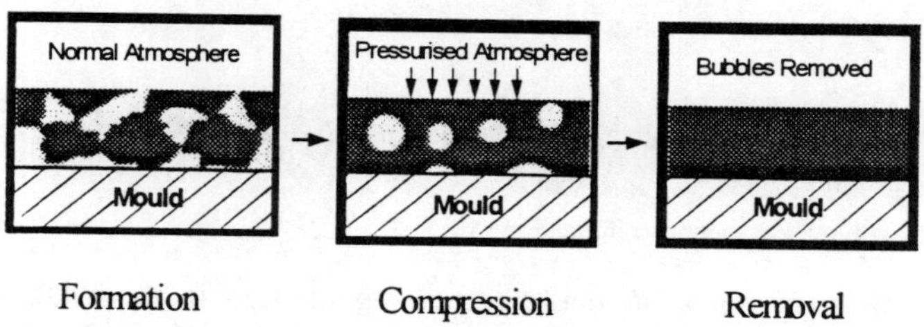

FIGURE 8 Pressurisation process

The effect of pressurisation was initially investigated[1] with the aid of a sealed test rig and a hot plate. During a series of tests, a polymer powder (MFI 3.2g/10mins, density 934 kg/m^3) was melted on a 300°C hot plate in pressurised atmospheres ranging from 0 to 6 bar (gauge pressure). Each test was observed using a video camera and recorded for further analysis. From these tests, it was observed that there was little or no difference experienced in the diffusion rates of bubbles in atmospheric pressure, compared to the diffusion of bubbles in a pressurised atmosphere. For both sets of conditions (i.e. pressurised and unpressurised) bubbles take between 15 and 25 minutes to diffuse at a relatively high melt temperature (190°C).

The reason why these initial pressure tests had no effect on bubbles or pores became apparent after considering the formation of bubbles in an unpressurised atmosphere. When bubbles form under normal conditions, they occur due to air being trapped between powder particles as they join together at their points of contact and then melt completely. The air trapped inside each bubble is equal in pressure to the air above the melt (atmospheric pressure). Therefore, there is no force exerted on the bubble by the surrounding atmosphere. During the initial pressure tests, the polymer was melted in a pressurised atmosphere; this resulted in the pressure inside the bubbles being equal to the pressure of the surrounding atmosphere. Therefore, like the test conducted in a normal atmosphere, the external pressure exerted no additional compressive force on the bubbles.

For mould pressurisation to be effective, a pressure difference has to be created between the pressure inside the bubble and the pressure exerted by the surrounding melt. This is achieved by first allowing the polymer to melt under atmospheric conditions. Then secondly, when the powder has completely melted, the pressure should be applied. This method of pressure application was investigated[1] using a new series of hot plate tests. During these tests, the effect of 0.5, 1, 1.5, and 2 bar (7.25, 14.5, 21.75 & 29 psi) pressure on the diameters of bubbles in the polymer melt (at 190°C) was investigated. Figure 9 illustrates how the diameters of bubbles which formed under atmospheric pressure conditions were influenced by increasing applied pressure levels. It can be seen that pressure, if applied after the polymer has melted, can successfully remove bubbles from within the polymer melt in a matter of seconds. Increasing the applied pressure level increases the rate at which bubbles are removed.

The effect of pressure on the surface porosity of the part was also studied[1] using the hot plate rig. During this series of tests the rig/melt interface was heated to 200°C and then the pressure was applied for a set period of time, for example, 1 bar (14.5 psi) for 10 seconds, then relieved and opened to atmosphere again.

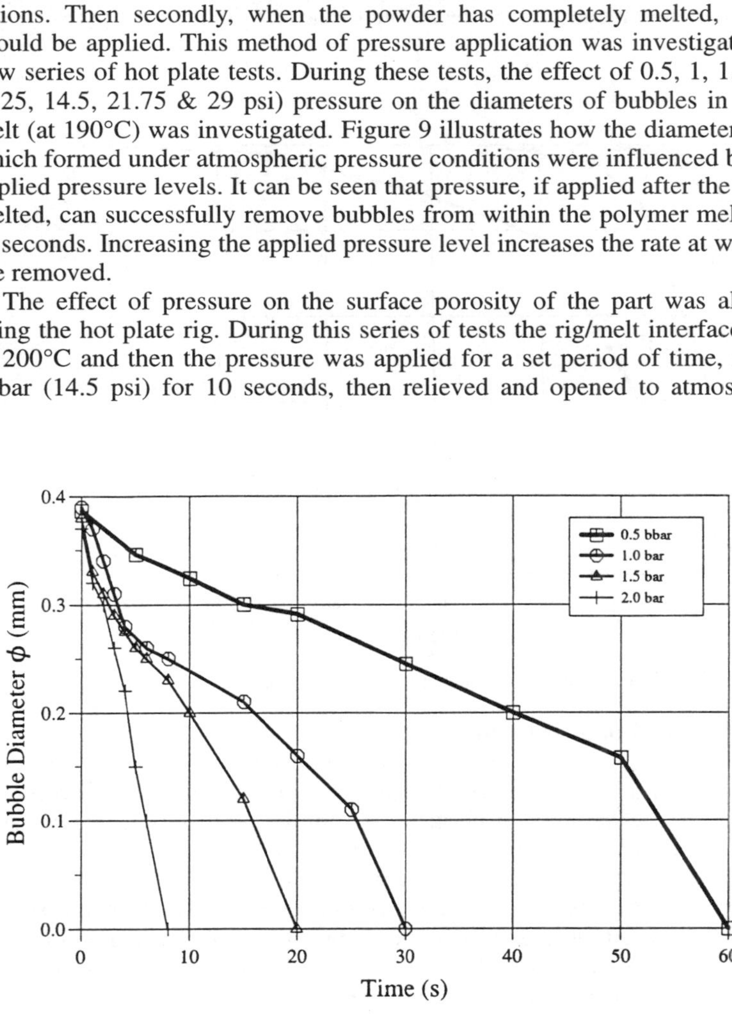

FIGURE 9 Effect of pressure on bubble diameter

The polymer was then cooled and the number of surface pores contained in a 10mm square area were counted. From these tests it was found that the application of pressure, even for a short period of time, reduced surface porosity significantly. For example, 1 bar pressure applied for 30 seconds removed the majority of surface pores. The overall effect of a range of pressures (0.5, 1, 1.5 and 2 bars) on surface porosity was determined, and is illustrated in Figure 10. It can be seen that pressure removes surface porosity in a similar fashion to the removal of pin holes contained within the melt. It appears that pressure removes the smallest surface pores instantaneously, and then gradually forces the air out of the remaining larger pores.

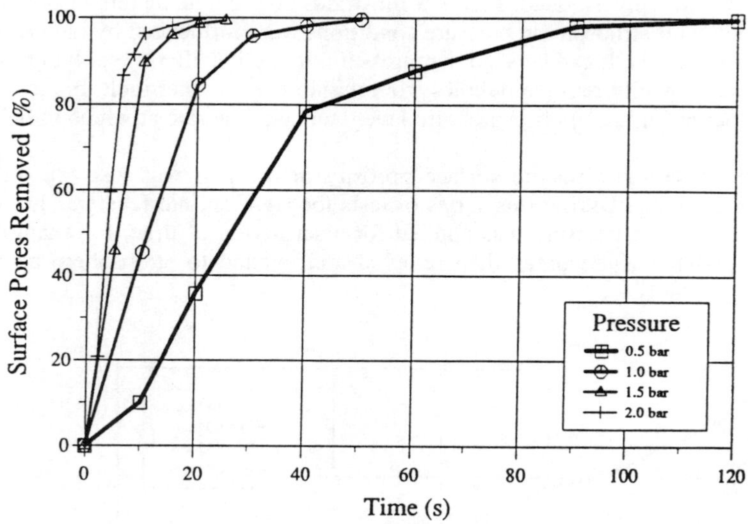

FIGURE 10 Effect of pressure on surface porosity

5.1 Pressurised Rotational Moulding

The effects previously mentioned were all observed using hot plate trials, but were subsequently shown to work equally well during rotational moulding. Rotational moulding trials on a wide range of plastics have confirmed that moderate pressure (0.5bar / 7.25psi) could successfully remove bubbles and pores, if applied after the polymer has melted. By using the gas pipeline which exists on the arm of many industrial machines, the mould can be pressurised (see Figure 11). Mould pressurisation thus provides a simple mechanical means by which bubbles and pores could be removed from the product. In addition, there are a number of other effects experienced when pressurising the mould, which can be used to the advantage of the moulder.

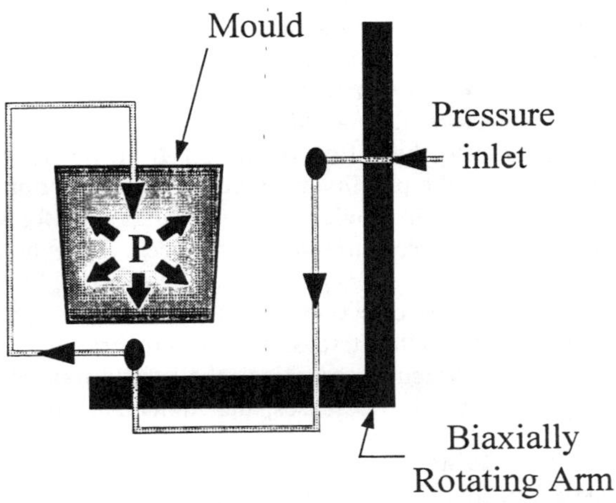

FIGURE 11 Mould pressurisation

It has been established[12] that under normal moulding conditions the optimum point to remove a polyethylene moulding from the oven for cooling is when its internal air temperature reaches 190°C-200°C. Impact tests have shown that at this peak internal air temperature the optimum mechanical properties of the moulding are achieved. It was felt that this offered the best compromise between leaving the plastic in the oven until all the pin holes had disappeared naturally and removing it before the thermal degradation effects became pronounced. As pressure has now been shown to remove bubbles and pin holes mechanically rather than thermally, the intriguing question arose: could the pressure be used to remove the bubbles much earlier than normal and thus facilitate earlier removal of the moulding from the oven?

A series of rotational moulding trials was set up to study the effectiveness of pressure to remove bubbles and also to improve mechanical properties. During these tests the oven temperature was set at 300°C and the mould was removed from the oven at a range of internal air temperatures (120, 135, 145, 160 & 180°C). When removing the mould from the oven, a 0.5 bar (7.25 psi) pressure was applied through the arm of the machine for a range of times (0, 1, 2, 5, 10 & 15mins). Table 2 outlines the test parameters, together with the effectiveness of pressure to remove bubbles from a low MFI material (3.2g/10mins) for a series of mouldings coded PT01 to PT18 (PT denoting Pressure Test).

During the rotational moulding trials it was noticed that the onset of pressure produced a sudden rise in the internal temperature of the mould. There was also a corresponding drop in temperature inside the mould when the pressure was released. This observation can be seen by comparing the Rotolog trace presented in Figure 12 when no pressure was applied, to Figure 13, when 0.5 bar had been applied at 180°C for 5 minutes. It was found that this rise in temperature inside the mould was a result of the gas pipeline mounted on the arm of the machine. During the heating stage of the moulding cycle, the air contained in the copper pipeline was being heated to the same temperature as the oven. The air in this pipeline was then used to pressurise the mould. Therefore, the mould experienced a rise in temperature with the onset of pressure due to the influx of hot air from the copper pipeline. When the pressure was released, the pipeline provided a means for hot air inside the mould to escape. This observation provided a useful indicator for the pressurisation process, identifying when the pressure was introduced, and when it was relieved.

The data in Table 2 illustrates clearly the effectiveness of pressure to remove bubbles - in particular, the effect increases with increasing melt temperature. This can be related to the polymer viscosity which decreases as the melt temperature increases. Therefore, as the polymer viscosity decreases, the molecular structure

Table 2
Effect of 0.5 bar (7 psi) Pressure applied at Various Internal Temperatures

Code	Mould Internal Temperature (°C)		Duration of Pressure (mins)	Approximate Percentage of Bubbles Removed
	Removed at	Peaked at		
PT01	125	148	0	0
PT02	125	152	1	10
PT03	125	154	2	60
PT04	125	153	5	80
PT05	125	154	10	90
PT06	125	155	15	95
PT07	145	160	0	0
PT08	145	166	1	50
PT09	145	167	2	90
PT10	145	167	5	95
PT11	145	168	10	100
PT12	145	166	15	100
PT13	180	185	0	0
PT14	180	190	1	70
PT15	180	188	2	95
PT16	180	189	5	97.5
PT17	180	190	10	100
PT18	180	187	15	100

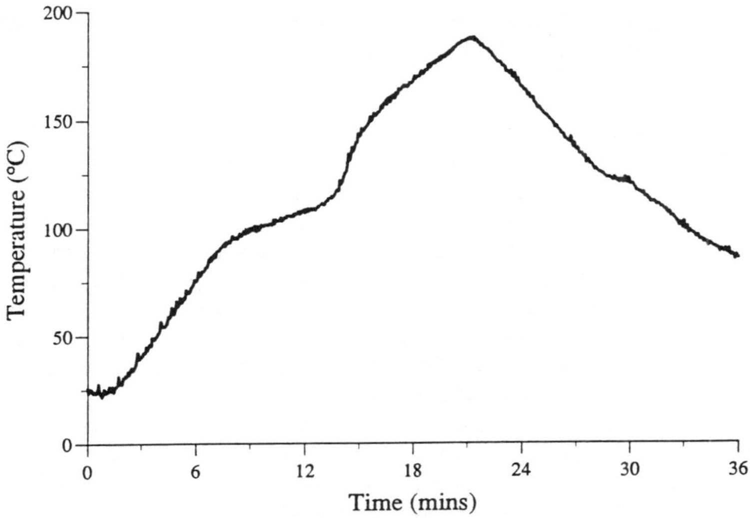

FIGURE 12 Normal Rotolog trace

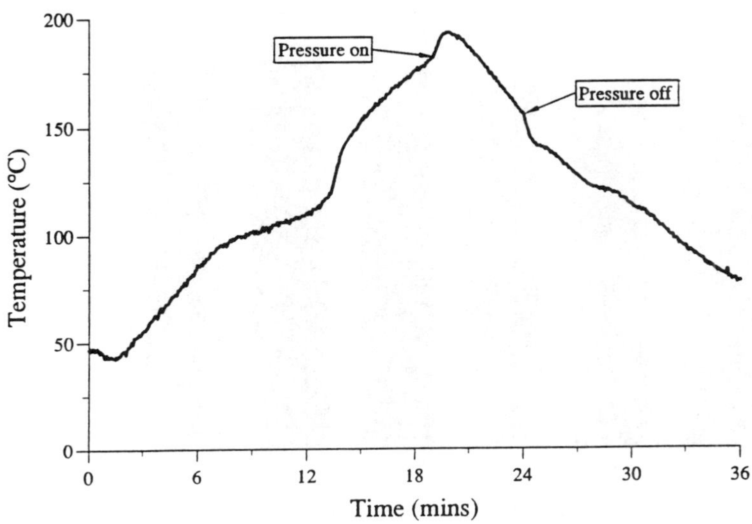

FIGURE 13 Rotolog trace with pressure applied at 180°C for 5 mins

weakens, providing an easier passage for air molecules to diffuse through the melt. For the series of mouldings coded PT01 to PT18 some mechanical properties such as impact strength and tensile modulus were measured to see if the use of pressure offered any advantages in this regard. Preliminary trials showed that there was no significant change in flexural properties.

Mechanical Properties

During the above investigation it was considered important to know the effects on mechanical properties. Therefore, low temperature (-20°C) impact tests were carried out on samples from each moulding. From the data generated, a trend was identified which indicated that mechanical properties increased when bubbles and pores were removed from the part. Figure 14 illustrates a typical trend of results for the series (PT07 to PT12) of mouldings removed from the oven at 145°C. With no pressure applied, the bubble count is 100%, which corresponds to the least energy that is required to break the sample. Bubbles weaken the polymer structure by providing crack initiation zones. When bubbles are removed, the impact strength typically rises by about 15%-20%, and on some occasions by as much as 25%.

Tensile tests were also performed on a series of samples from each moulding. Again the advantages of using pressure were highlighted. The tensile tests indicated that the removal of bubbles increases both the tensile strength (by approximately 5%) and the elongation, before the sample breaks. Bubbles in the tensile samples cause weaknesses in the polymer structure, through which cracks form. These

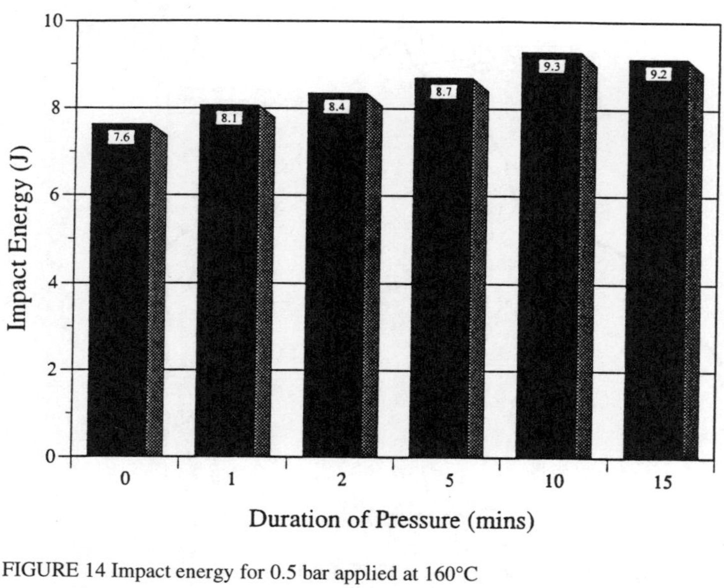

FIGURE 14 Impact energy for 0.5 bar applied at 160°C

weaknesses can be easily seen during testing and are illustrated at xl0 magnification in Figure 15. During the tensile test, the bubbles elongate, increasing in size, and eventually they cause the sample to break. If pressure is used to remove bubbles, as is the case for Figure 16, then the sample has no points of weakness. In this case elongation is continuous and the sample does not break.

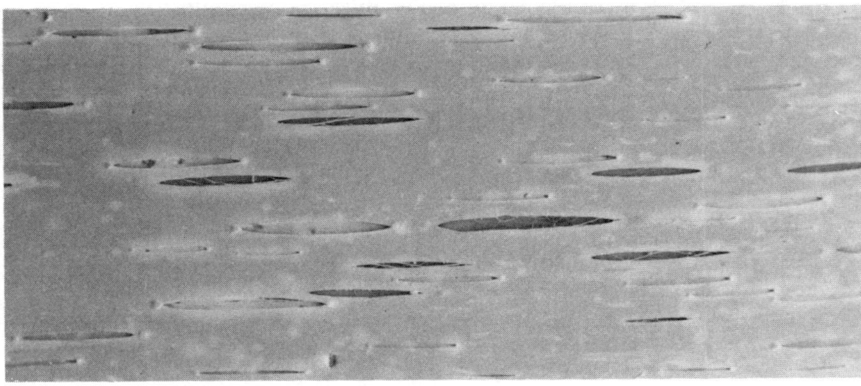

FIGURE 15 Tensile sample from moulding coded PT01 (magnification x10)

FIGURE 16 Tensile sample from moulding coded PT12 (magnification x10)

Cycle Time

From the mechanical test results it can be seen that, with the application of pressure, the mould can be removed from the oven at a lower temperature with no adverse effect on the properties of the part. Indeed, part properties increase, as the weaknesses incurred due to bubbles and pores have been removed. For example, moulding PT06 which was removed from the oven at 125°C (with pressure), with no bubbles or pores, has better impact and tensile properties than moulding PT13 (without pressure), which was removed at 180°C. This means that not only does pressure increase the mechanical properties of the product, but it also reduces the manufacturing cycle time to produce the moulding. The Rotolog trace shown in Figure 17 illustrates how the cycle time is reduced when the mould is removed from the oven at a lower temperature. For this particular illustration, 8 minutes could be saved during the rotational moulding cycle if the mould was removed from the oven at 135°C. If the moulding was de-moulded at 100°C then 8 minutes would represent a 20% saving on the rotational moulding cycle time. Removing the mould from the oven at a lower temperature also means that the polymer will be less degraded, as it will have been subjected to a lower temperature for a shorter period of time.

Pressure Level

The effect of the applied pressure level on the mechanical properties of the product was also investigated[1]. During this series of tests, the polymer was heated to 190°C and then removed from the oven. As it was being removed from the oven,

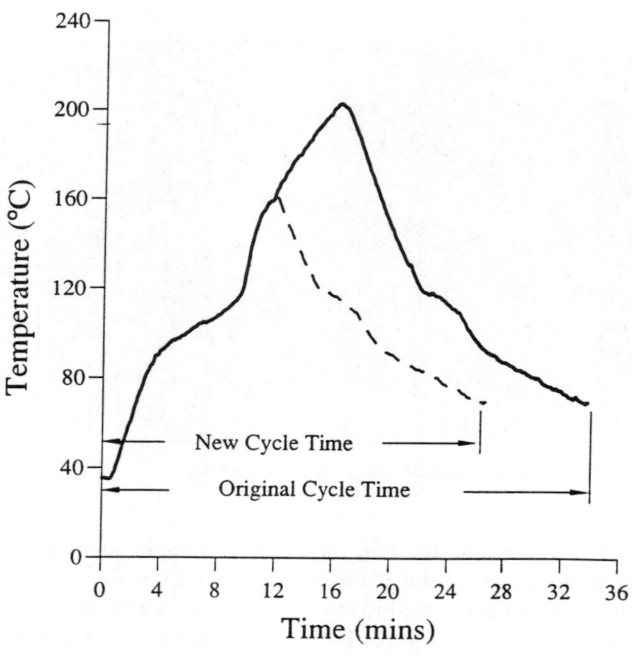

FIGURE 17 Rotolog trace illustrating the possible reduction in cycle time due to pressure

pressure levels ranging from 0-3 bar (0-43.5 psi) were applied for 10 minutes and then relieved. Following rotational moulding, some mechanical properties were investigated (impact and tensile).

From the impact test results (-20°C), it may be seen that all three applied pressure levels removed bubbles and pores completely, and hence increased the toughness of the mouldings. However, there was no noticeable increase in impact strength experienced for each increasing pressure level. But the tensile properties were affected by the applied pressure level (see Table 3). As the applied pressure level increased, the tensile strength and tensile modulus both increased. The tensile properties improved because increasing the pressure level consolidates the polymer structure and improves its resistance to the externally applied forces.

Table 3
Tensile Data for Increasing Pressure Levels

Applied Pressure (bar)	Tensile Strength (MN/m^2)	2% Secant Modulus (GN/m^2)	Elongation at Break (%)
0	17.53	0.323	200-500
1	18.05	0.327	1000+
2	18.17	0.333	1000+
3	18.76	0.347	1000+

Product Thickness

The effectiveness of pressure to remove bubbles and pores from thick walled polyethylene mouldings has also been gauged[1]. During these trials, 6mm and 9mm thick parts were rotationally moulded. Mould pressurisation (0.5bar/ 7.25psi) was again shown to remove bubbles and pores very effectively. Therefore, it was concluded that the action of mould pressurisation to remove bubbles was independent of part wall thickness, at least for thicknesses in the range of 1 to 9 mm and there is no reason to suspect that it will not also work for greater wall thicknesses.

Moulding Material

The application of pressure to the polymer melt has also been shown to remove bubbles and pores from other materials such as Polypropylene, Nylon 12, EVA, EBA and Polycarbonate. Figures 18 and 19 illustrate the dramatic effect achieved with EBA, a material which is notoriously bad in regard to bubbles. Impact tests showed that the application of pressure has less beneficial effect on the impact strength of these materials (compared with polyethylene). However, it should be noted that the aesthetic properties of some of these materials are just as important as their mechanical properties. For example, with polycarbonate its transparency is

FIGURE 18 EBA with pressure applied

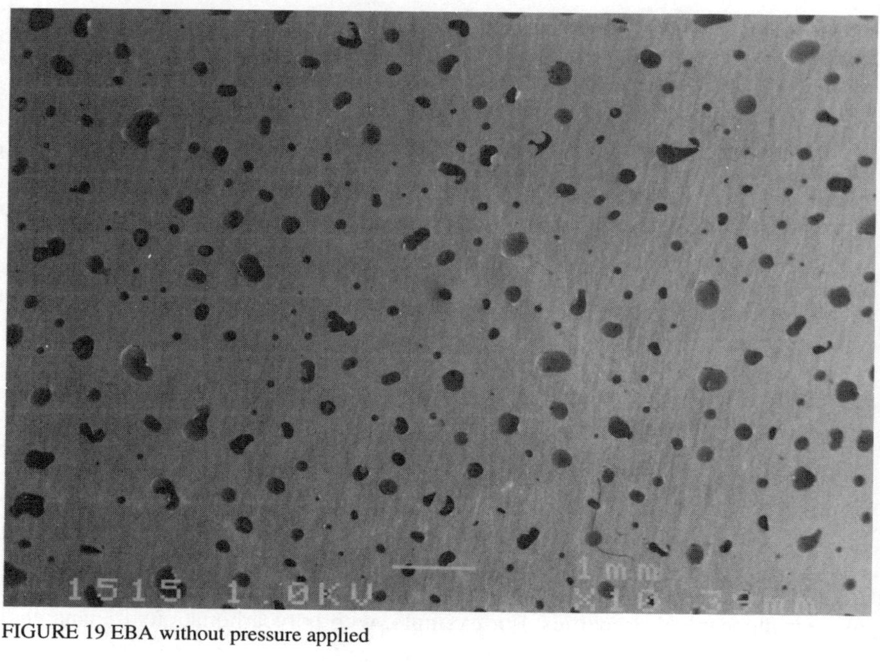

FIGURE 19 EBA without pressure applied

critical in some applications. For materials like Nylon, which is used to manufacture chemical storage tanks, the removal of voids and pores reduces the risk of possible leakage.

5.2 Using Pressure as a Blowing Agent

Blowing agents are additives used during rotational moulding to produce cellular materials, that is, materials containing large proportions of fine cells filled with gas. Their purpose is to increase the product stiffness per unit weight with minimal additional cost of materials. It was postulated that if the method of the application of pressure was reversed, that is, by applying pressure before the polymer melted and releasing it after the polymer melted, then pressure could be used as a blowing agent. To test this idea, a series of rotational moulding trials was carried out, and these confirmed that pressure could be used to increase the wall thickness of the part. In the first trial, the material was moulded normally without pressure. In the second trial, 2 bar (29 psi) pressure was maintained during the heating cycle and released when the polymer had melted. Figures 20 and 21 illustrate cross-sections at xl0 magnification taken from each moulded item. From the micrographs it can be seen that, with no pressure used, bubbles are small and the moulded sample is approximately 6mm thick. When pressure was introduced before the polymer had melted, the bubbles which formed were pressurised. When the pressure acting on the melt was relieved, then the bubbles expand to new diameters with their internal pressure equalising to the pressure of the surrounding melt. This effectively increases the parts thickness, without any additional material being added. Figure 21 illustrates a 30% increase in part thickness.

5.3 The Use of a Vacuum to Remove Bubbles

It has been previously outlined that pressurisation of the polymer after it has melted successfully removes bubbles. Alternatively, a vacuum can be used to provide a pressure difference between the bubble and the surrounding atmosphere, and so aid bubble removal. If a vacuum is applied before heating and rotation commence, then it will be found that the number of voids which form are significantly reduced because the majority of air molecules have been removed. However, the bubbles which do form tend to increase in size as the polymer temperature increases. This is because the air pressure inside the bubble rises due to the surrounding melt temperature, and the bubble is able to expand with greater freedom as there is no atmosphere above the melt. However, if the vacuum is released once all the melt is formed, then it will be found that all the bubbles disappear instantly from the melt. This is because, effectively, the introduction of atmospheric pressure caused a pressure difference between the pressure inside and outside the bubble and this pressure difference caused the void to collapse.

6. EFFECT OF PRESSURE AND VACUUM ON THE MOULD

It should be noted that the use of pressure and vacuum have certain limitations. Mould pressurisation can be extremely dangerous if care is not taken. It is

FIGURE 20 Cross-section of a moulding without pressure (magnification x10)

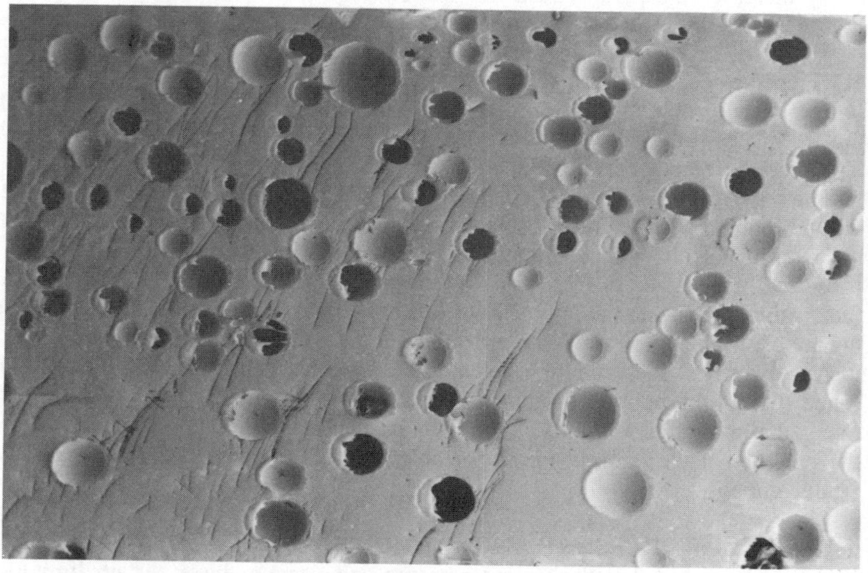

FIGURE 21 Cross-section for 2 bar introduced before the polymer has melted and relieved after the polymer has melted (magnification x10)

recommended that pressure should only be used for moulds which are sufficiently strong to withstand the forces involved. This method for removing bubbles is not suitable for large moulds fabricated from sheet steel, as moderate pressures will cause distortion or even mould fracture.

In general, it is likely the use of pressure is not suited to many of the moulds used in the industry today. This is because their clamping is not sufficiently good to withstand the internal forces caused by pressure. When manufacturing a new mould specifically for use with pressure, the following suggestions are made:

(a) Aluminium moulds need to be three times thicker than mild steel moulds to provide equal strength.
(b) Flat sections on moulds should be avoided. The design of the mould should incorporate strength features where possible.
(c) When using pressure, the quality of the moulding is dependent on the surface finish of the mould - as pressure provides better part definition.
(d) Pressure will tend to force the molten polymer into the parting line. Therefore, the mould parting line should be machined if possible, and contain an 'O' ring.
(e) The mould, or the pressure connection on the mould, should incorporate a pressure relief valve.

It should also be noted that not all moulds would lend themselves to the application of a vacuum. To facilitate this, moulds would require an 'O' ring for sealing purposes, otherwise suction holes would be created at the parting line. Also, moulds would need to be strong enough to withstand the forces caused by the applied vacuum, which are much greater than the forces exerted by a small positive pressure (5 psi).

When these design features are included, the moulder is in a position to utilise "Roto-blow" moulding. This is a hybrid of rotational moulding and blow moulding where the moulder is in a position to benefit from the better material distribution in the former method and better product quality in the latter.

REFERENCES

[1] Spence, AG "Analysis of Bubble Formation and Removal in Rotationally Moulded Products" PhD Thesis, Queen's University of Belfast, 1994.

[2] Rao, MA and Throne, JL "Theory of Rotational Molding - Part III: Sinter-Melting and Degradation", The Society of Plastics Engineers, pp759-761, May 1972.

[3] Progelhof, RC, Cellier, G and Throne, JL "New Technology in Rotational Moulding: Powder Densification", Society of Plastics Engineers, Annual National Technical Engineering Conference, p762, 1982.

[4] Kelly, PY "A Microscopic Examination of Rotomoulded Polyethylene", Du Pont Canada Incorporation, Unpublished Report.

[5] Crawford, RJ and Scott, JA "The Formation and Removal of Gas Bubbles in a Rotational Moulding Grade of Polyethylene", Plastics and Rubber Processing and Applications, pp85-89, 1987, Vol.7.

[6] Crawford, RJ and Xu, L "Analysis of the Formation and Removal of Gas Bubbles in Rotationally Moulded Thermoplastics", Material Science, pp2067-2074, 1993, Vol.28.

[7] Nelkon, M and Parker, P "Advanced Level Physics", London: Heinemann Educational Books, 1977.

[8] Barnes, HA, Hutton, JF and Walters, K "An Introduction to Rheology", Oxford: Elsevier, 1989.
[9] Heath, RJ "A Review of the Surface Coating of Polymeric Substrates", Progress in Rubber and Plastics Technology, pp369-401, 1990, Vol.6.
[10] "Metals Handbook 9th Edition" Ohio: American Society for Metals, 1985.
[11] Evans, UR "Corrosion and Oxidation of Metals", London: Edward Arnold Ltd, 1960.
[12] Crawford, RJ and Nugent, PJ "Impact Strength of Rotationally Moulded Polyethylenc Articles", Plastics, Rubber & Composites Processing & Applications, pp33-41, 1992, Vol. 17.

ACKNOWLEDGEMENTS

The authors are grateful to the Science and Engineering Research Council, Lin Pac Rotational Moulders and Neste Chemicals (now Borealis) for the financial support of this work. The work benefited greatly by the regular inputs from the staff at the sponsoring companies and from Dr Bob Pittillo of the Polymer Engineering Group. The authors are also indebted to Dr Jovita Oliveira and Dr Jose Corvas from the Universidade de Minho, Portugal for assistance with the bubble analysis and measurement of melt rheology at low shear rates. Thanks are also due to Dr Steve Holding from Rapra Technology for determining the molecular weight distributions of various materials.

DISCLAIMER

The effects of pressure and vacuum described in this chapter are known to be safe when used correctly. However, it is up to the moulder to check that moulds are sufficiently strong to withstand the forces which arise when pressure or vacuum are applied. Great care should be taken to ensure that excessive forces are not generated in the mould.

CHAPTER 11
Rotational Moulding of Liquid Polymers

E. Harkin-Jones and R.J. Crawford

1. INTRODUCTION

Liquid polymers offer many advantages over powders to the rotational moulding industry (see Table 1). Not surprisingly therefore, a number of liquid systems have been in use in the industry over the years. PVC plastisols were the earliest liquid polymers to be used in the rotational moulding process, and today account for approximately 3% of Western European plastisol consumption. Caprolactam based reactive liquid systems have been used for making large fuel storage tanks since 1959[1-3], when the 'Unicast' process was developed by the Allied Chemicals Corporation. Thermosetting liquid polymers have been in use since the early 1970's in applications ranging from small instrument housings to furniture. In 1976 a patent was granted to Monster Mouldings Ltd in Canada[4] for the 'process moulding of polymers at ambient temperature'. This patent covered polyurethane, polyester, epoxide, silicone, phenolic and urea formaldehyde systems.

Although there is this long history of the use of liquid polymers in the rotational moulding industry, and the potential advantages of their use in the process are many, the number of moulders actually using liquid polymers is small. The main obstacle to the development of liquid polymers as rotational moulding materials is in the many perceived difficulties in processing them. The main problems reported relate to an uneven distribution of material on the mould walls and surface bubbles in the finished part[5]. These problems, coupled with a lack of research effort in this area, have been sufficient in the past to limit the use of liquid polymers in the rotational moulding industry.

In recent years, however, commercial interest in the rotational moulding of liquid polymers has been revived and current research efforts in this area have focused on demystifying the processing of such materials and providing processing guidelines to the industry[6].

2. MATERIALS

Liquid polymers, for the purposes of this chapter, are classified into two areas:

(i) PVC plastisols where the liquid is a suspension of PVC resin particles in a plasticiser. When the plastisol is heated, the plasticiser diffuses into the resin

Table 1 Advantages to be gained when moulding liquid polymers

1. In general, liquid polymers are processed at much lower temperatures than those required for powders (PVC plastisols are the exception). A number of polyurethanes and epoxides can be processed at room temperature.
2. It is possible to employ a greater range of mould materials. With many liquid polymers heating and cooling of moulds is not required and severe thermal cycling of the mould tool is avoided. Ambient temperature mouldings can be made in expensive, light, glass fibre moulds.
3. Cycle times can be as short as two minutes for materials such as reactive liquid nylon. Cycle times are also generally independent of part size/wall thickness.
4. Liquid polymers give excellent reproduction of surface detail, threaded inserts etc.
5. Because lower mould temperatures are generally required and exothermic heats of reaction may be used to heat the mould for the next cycle, the processing of certain liquid polymers is very energy efficient.
6. Capital equipment costs can be dramatically reduced for ambient temperature moulding since only a drive system is required to provide the biaxial rotation.
7. Liquids offer greater possibilities for reinforcing than do powders. With powders the reinforcing fibres, being larger than the powder particle, tend to form a coating on the inside of the part with very little distribution throughout the part wall section. Liquids allow the possibility of premixing with fibres prior to mould filling. If the initial viscosity is not too low and the cure time is short then the fibres will be distributed throughout the section.

particles and the system increases in viscosity to become a dry coherent mass. On further heating, fusion of the polymer molecules takes place; and upon cooling, the material exhibits the intrinsic strength of a plasticised PVC.

(ii) Reactive liquid polymers such as polyurethanes and nylon. These systems consist of two or more components which, when mixed together and heated (some systems do not require the addition of heat), will react chemically causing an increase in liquid viscosity and will eventually form a solid. Cooling is not generally required in such systems but may be employed to reduce handling temperatures.

The three most important characteristics of a rotational moulding grade of liquid polymer are: (i) initial viscosity (ii) minimum viscosity during processing and (iii) the rate of viscosity increase as the liquid 'cures' to form a solid. These characteristics will determine the mould coverage, bubble content, wall thickness distribution and cycle time of moulded parts. The significance of these characteristics in determining the moulding behaviour of various liquid polymers is considered in the following section.

3. FLOW BEHAVIOUR OF LIQUID POLYMERS IN A ROTATING MOULD

3.1 General Flow Behaviour

The liquid polymer initially forms a pool at the base of the mould. As rotation commences, a layer of liquid is lifted by the rotating wall. The depth of this layer depends on the linear velocity of the mould (i.e. the product of angular rotation speed and mould diameter) and on the viscosity and density of the liquid[7]. The greater the viscosity the thicker the layer of liquid that may be picked up by the mould wall as it rotates out of the liquid pool. Similarly, the greater the velocity of the mould wall, the greater the layer of liquid it can pick up. Layer thickness also increases with decreasing liquid density.

As the liquid heats up, some re-pooling of the liquid adhering to the mould walls may occur. This is due to a decrease in the viscosity of the liquid on heating. This re-pooling may lead to bubble formation if the liquid begins to drop from the highest point of the mould into the pool (thus entrapping air). The extent of re-pooling is determined by the temperature dependence of the material's viscosity, the initial layer thickness on the mould wall and on the point at which the liquid begins to 'react' and hence increase in viscosity.

Once the liquid begins to increase in viscosity (due to a chemical reaction in nylon and polyurethane and due to diffusion of plasticiser into the resin in PVC plastisols), the layer of liquid that can be supported by the rotating wall increases. At some point in time, the pool of liquid is entirely dissipated onto the rotating mould walls.

3.2 The Effect of Initial Viscosity and Minimum Viscosity on Flow Behaviour

If the initial liquid viscosity is low then only a thin layer of liquid is picked up; if the viscosity is high then a thicker layer will be picked up to coat the mould wall. If the initial viscosity of the liquid is too high then the part of the mould to first come into contact with the liquid pool will lift the entire pool and the material may spread in an uncontrolled fashion over the mould wall or may remain on that part of the mould which is initially contacted. Either way a faulty moulding will be produced if the material cures in that particular state. Bubble formation is also greatly increased in this case since large drops of material detach from the mould wall as it reaches an overhead position. These drops entrap air as they fall back onto the base of the mould. The problems occurring due to too high an initial viscosity may be alleviated if the liquid exhibits a significant reduction in viscosity as it heats up. This reduction in viscosity will allow the liquid to flow out from the heavy layer initially picked up by one part of the mould wall and this liquid may then coat the walls in a more uniform fashion.

Too low an initial viscosity will also lead to problems. Bubble formation becomes a greater problem in this case due to the low viscosity liquid sloshing around in the rotating mould. Rotation speeds must be kept to a minimum in such cases.

3.3 The Effect of Rate of Increase of Viscosity on the Flow Behaviour of Liquids

The rate of increase of liquid viscosity during processing has a very significant influence on the state of the finished part. If viscosity increases too rapidly then the liquid will not have sufficient time to coat the mould walls evenly, and a part with non-uniform wall thickness will result. The viscosity should increase gradually in order to allow the pool of liquid time to spread uniformly onto the mould walls. It should be noted, however, that if the initial viscosity of the liquid is too high, resulting in the behaviour mentioned earlier, then the rate of viscosity increase has little or no effect on the outcome of the finished moulding.

Tables 2 and 3 show the standard deviation from the mean of the wall thickness measurements of a rotationally moulded 300 mm cube, 3 mm nominal wall thickness, made using a number of different liquid polymers. Table 4 gives details of the individual materials.

At both rotation speeds, Nyrim has the worst part thickness uniformity while Hyperlast 7850506 has the best. Reference to Figures 1 and 2 provides some

Table 2 Comparison of part wall thickness uniformity for various grades of liquid polymer rotationally moulded at 5.8 rpm

Material	Standard deviation of part wall thickness from the mean.(mm)
Nyrim	0.75
Hyperlast 7853184	0.63
Hydro PVC	0.45
Acrol PVC	0.46
Hyperlast 7850506 @ 45°C mould temperature	0.40
Hyperlast 7850506 @ 70°C mould temperature	0.36

Table 3 Comparison of part wall thickness uniformity for various grades of liquid polymer rotationally moulded at 13.3 rpm

Material	Standard deviation of part wall thickness from the mean (mm)
Nyrim	0.64
Hydro PVC	0.43
Acrol PVC	0.43
Hyperlast 7850506 @ 45°C	0.38
Hyperlast 7853184 @ 80°C	0.28
Hyperlast 7850506 @ 70°C	0.27

Table 4 Details of materials referred to in Table 1

Material Code Name	Material Type	Supplier
Nyrim	Nylon block copolymer	DSM Rim Nylon, The Netherlands
Hyperlast 7853184	Polyurethane	Mac Pherson Polymers, England
Hyperlast 7850506	Polyurethane	Mac Pherson Polymers, England
Acrol PVC	PVC plastisol (D.O.P)	Acrol Ltd., England
Hydro PVC	PVC plastisol (D.O.P.)	Hydro Coatings, England

insight into the reasons for the different moulding behaviours of these materials. Figures 1 and 2 show the viscosity changes in each material during the moulding process. It can be seen that Nyrim has the lowest viscosity of any of the materials tested. Even when re-pooling due to heating occurs in the other materials, the reduced viscosity is never as low as that of Nyrim. Having a low initial or minimum viscosity is in itself not bad if low rotation speeds are employed to prevent sloshing of the liquid in the mould. However, when a low initial viscosity is combined with a rapidly increasing viscosity during cure, as is the case for Nyrim, the result is a moulding with poor uniformity of wall thickness.

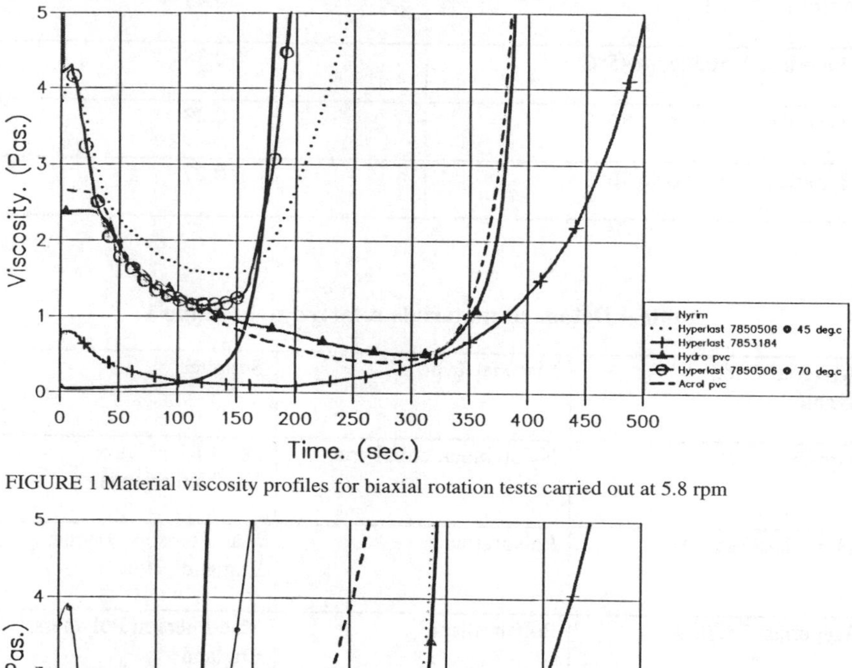

FIGURE 1 Material viscosity profiles for biaxial rotation tests carried out at 5.8 rpm

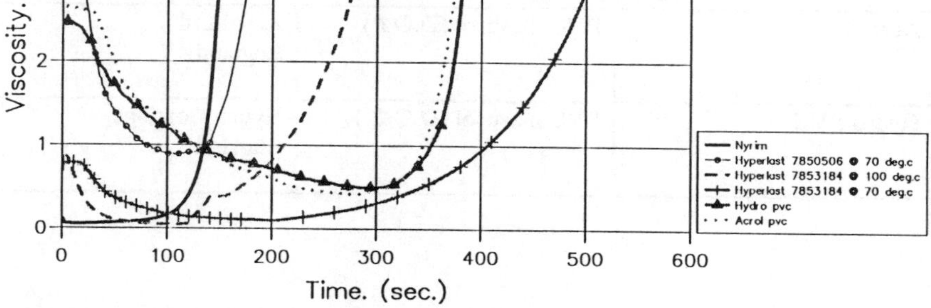

FIGURE 2 Material viscosity profiles for biaxial rotation tests carried out at 13.3 rpm

Hyperlast 7853184 has the next lowest initial viscosity. It is next in line to Nyrim in terms of poor uniformity of wall thickness at rotation speeds of 5.8 rpm. Its moulding behaviour improves dramatically, however, as the rotational speed increases. This improvement in part uniformity with increasing rotation speed is more apparent for Hyperlast 7853184 than for Nyrim due to the fact that the polyurethane has a higher initial viscosity. Increasing speed therefore allows the mould to lift a greater liquid layer thickness than is the case with Nyrim. The amount of liquid remaining in the pool when the viscosity begins to increase rapidly will thus be less for Hyperlast 7853184 and the final part uniformity will be better than that for Nyrim.

The PVC plastisols have viscosity profiles that increase as rapidly as Nyrim and more rapidly than Hyperlast 7853184 yet they produce better parts than both of these materials. This can be attributed to the fact that the minimum viscosity attained by the plastisols during the moulding process is never as low as that for Nyrim and Hyperlast 7853184. The plastisols will not therefore have a large pool of liquid to dissipate onto the mould walls once the viscosity begins to increase since most of the liquid pool has been lifted onto the mould wall early in the process. This will result in a more uniform part wall thickness.

The Hyperlast 7850506 mouldings have the best part wall thickness uniformity of the materials under consideration. This is due to a combination of two factors: (1) a high initial viscosity and a small degree of re-pooling, which means that most of the liquid pool will coat the mould walls after the first few rotations of the mould and, also, very little material will re-pool due to heating; (2) a relatively slowly increasing viscosity profile which allows the remaining pool to be gradually lifted by the rotating wall to allow a uniform coating of the mould.

It must be noted that Hyperlast 7850506 does not perform as well at a mould temperature of 45°C as it does at 70°C. Although it has a more gently increasing viscosity profile at 45°C its minimum viscosity during processing is never sufficiently low to allow the material picked up in a heavy layer at the start of rotation to spread out evenly over the mould walls. In fact, at a rotation speed of 13.3 rpm part of the mould will remain untouched by material altogether because the high initial liquid viscosity combined with a high rotation speed means that the pool is lifted by one part of the mould wall and tends to cure in this configuration. It is therefore necessary when moulding high viscosity materials such as Hyperlast 7850506 to use a low rotation speed and an initial mould temperature which is sufficiently high to allow re-pooling of the liquid which will then result in more even coverage of the mould.

It is obvious from the preceding information that not only is the rate at which viscosity increases during processing important but so also is the initial material viscosity and the minimum viscosity attained during processing. Another factor which must be considered when moulding liquids is the presence of bubbles in the finished part. Of the materials considered here, Nyrim tends to produce mouldings with the greatest number of bubbles while Hyperlast 7850506 has the least number. These are respectively the lowest and highest viscosity materials under consideration. Bubble formation is therefore a function of material viscosity. The higher the viscosity the more likely the liquid is to adhere to the mould wall as it approaches the highest point of rotation. There is thus less likelihood of the liquid dripping off the wall back into the pool, entrapping air as it falls and creating air bubbles in the pool. High viscosity liquids will also be less likely to slosh about in the mould. This is a problem with low viscosity liquids which again leads to bubble formation.

4. THE INFLUENCE OF PROCESSING CONDITIONS ON THE WALL THICKNESS UNIFORMITY AND ON BUBBLE CONTENT OF MOULDED PARTS

4.1 Rotation Speed

Figure 3 shows the influence of primary arm rotation speed on part wall thickness uniformity (standard deviation from mean) of mouldings made from Nyrim. The greater the rotation speed, the greater the thickness of liquid layer that can be picked up by the rotating wall. This ensures that only a small pool of liquid remains in the mould base to be lifted onto the walls once the liquid viscosity begins to increase rapidly. There appears to be little benefit in increasing the speed above 7 rpm, and lower rotation speeds should be encouraged in order to minimise build-up of material at mould corners (due to centrifugal effects) and also to minimise bubble formation (see Figure 4).

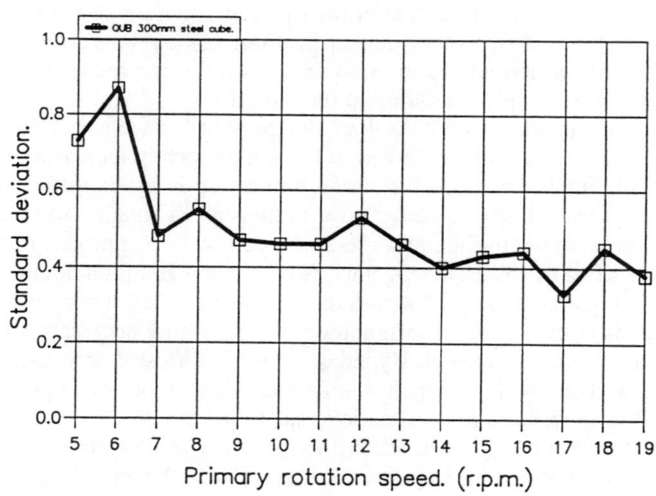

FIGURE 3 The effect of primary rotation speeds on Nyrim part wall thickness uniformity

4.2 Rotation Ratio

Optimum material distribution for a cube moulding made from Nyrim occurs at a rotation of 4.5:1. This is similar to the optimum ratio for symmetrical parts made with plastic powders.

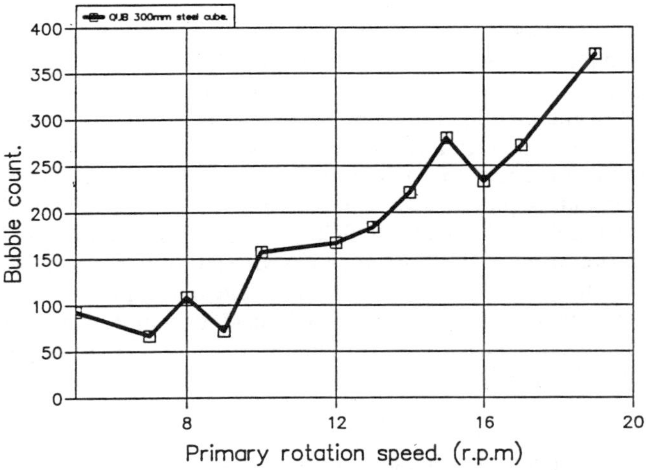

FIGURE 4 The effect of primary rotation speed on the bubble content of Nyrim parts

4.3 Shot Size

The effect of shot size (i.e. the quantity of liquid required to produce a moulding of a particular wall thickness) on the wall thickness uniformity of finished Nyrim parts is shown in Figure 5. Non-uniformity increases dramatically as shot size increases. This occurs because the greater the shot size the greater will be the size of liquid pool existing as the viscosity begins to increase rapidly. With a rapidly increasing liquid viscosity the liquid will not have sufficient time to coat the mould walls evenly prior to solidification. The rotating wall lifts a heavy layer of high

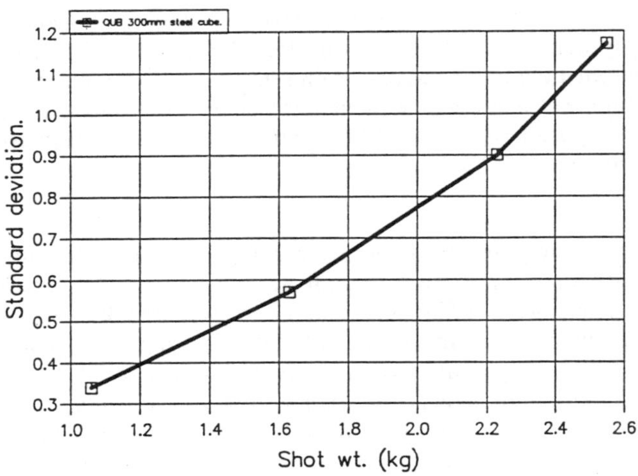

FIGURE 5 The effect of shot size on Nyrim part wall thickness uniformity

viscosity liquid which then spreads in an uncontrolled fashion over the mould surface and solidifies in this configuration to produce a bad moulding. Bubble formation in such parts is also very high since some of the heavy liquid layer falls from the highest point of the mould, entrapping air as it does so. In order to minimise bubble formation and non-uniform material distribution in thick walled mouldings, it is recommended that the moulding be carried out in two or more stages. In a two shot moulding process, for example, half the required quantity of liquid is placed in the mould and rotated until it solidifies. The rotation is then stopped and a second shot of material is added and processed. In this way the wall thickness uniformity of the finished part is improved and the bubble content is reduced. Note in Figure 5 that a 2.4 kg moulding will have its wall thickness standard deviation from the mean reduced from 1.05 mm to 0.4 mm if it is made by a two shot process using 1.2 kg per shot.

4.4 Initial Mould Temperature

Increasing initial mould temperature leads to a reduction in Nyrim part wall thickness uniformity. This occurs because the rate of increase of Nyrim viscosity is directly proportional to temperature, and the likelihood of achieving a uniform coating of the liquid pool onto the mould wall prior to solidification is reduced. It should be noted here that a slightly higher initial mould temperature may be beneficial for high viscosity materials such as Hyperlast 7850506 for the reasons stated in 3.3.

5. MOULDING EQUIPMENT

5.1 Material Handling

Single component liquid systems such as PVC plastisols are dispensed cold into the mould. These materials may be poured from a jug through a funnel into the mould or may be dispensed automatically from a machine. Two- or multi-component systems must be thoroughly mixed prior to dispensing into the mould and generally require sophisticated automatic mixing and dispensing machines. Materials such as Nyrim which are moisture sensitive must be maintained under a dry nitrogen blanket prior to mixing, and exposure of the liquid to the atmosphere during dispensing should be kept to a minimum.

5.2 Moulds

Since the temperature range required to produce mouldings from liquid resins is generally lower than that required for powders, a greater range of mould materials can be used (e.g. room temperature curing polyurethanes may be moulded in glass fibre or wood). Liquid polymers requiring a high initial mould temperature may be cast in any of the mould materials suitable for powders. Care must be taken, however, when using cast aluminium moulds due to the possibility of porosity. Liquids with a low initial viscosity will flow into the pores and solidify there. When the moulding is demoulded the solid polymer in the pores may dislodge aluminium around the edges of the pores thus increasing their size.

Due to the low viscosity of liquids compared with polymer melts, it is necessary to modify moulds to prevent leakage at the parting lines. This can be achieved by machining a channel in the mould flange and placing a flexible material (such as Viton rubber or high temperature silicone sealant) in the channel to provide a watertight seal when the mould is closed.

A factor in mould design which has been found to have a significant influence on the distribution of material in finished parts made from liquid polymers is the corner radius. When a liquid of low initial viscosity such as Nyrim is moulded in a mould with 90° corners, a part with poor distribution of material at the mould corners is obtained. The moulding is heavy at one side of the corner and light at the other side. In order to minimise this effect, radiussed corners must be employed, with the optimum radius being in the region of 25 mm for Nyrim mouldings.

Overall mould shape also has an important influence on the wall thickness uniformity and bubble content of finished parts. Mouldings produced in a smoothly contoured mould will result in a product with a more uniform wall thickness distribution and fewer bubbles than those produced in more angular moulds. In fact, mould shape has a greater influence on the quality of the finished part than any other factor when moulding liquids of low initial viscosity such as Nyrim (see Table 5). This improvement in wall thickness uniformity (and reduction in bubble content) can only be attributed to the fact that the initially low viscosity Nyrim liquid experiences a much less disturbed flow in the smoothly contoured mould. The flow of a low viscosity liquid in a more angular mould is more irregular, with waves being set up in the liquid as it sloshes about during rotation. These waves then tend to be frozen into the part when the liquid solidifies. This effect of mould shape on finished part quality is only valid for low viscosity liquids such as Nyrim. When a liquid of higher viscosity such as Hyperlast 7850506 is moulded in differently shaped moulds no significant change in part quality is observed (see Table 6). In this case pool disturbance is minimal due to the high viscosity of the liquid, and the likelihood of waves setting up is also less. This insensitivity of part quality to mould shape is obviously a desirable characteristic for a rotational moulding liquid.

Table 5 Comparison of methods of improving Nyrim moulding quality

Method of Improving Moulding Quality	Standard Deviation of Part Wall Thickness From The Mean (mm)
Cube mould and single shot @ 12 rpm (control)	0.41
Cube mould and double shot @ 12 rpm	0.33
Smoothly contoured mould and single shot @ 12 rpm	0.29

Table 6 The effect of mould shape on the uniformity of part wall thickness in Hyperlast 7850506 mouldings

Method of Improving Moulding Quality	Standard Deviation Of Part Wall Thickness From The Mean (mm)
Cube mould and single shot	0.27
Smoothly contoured mould and single shot	0.27

6. SOME RECOMMENDATIONS FOR THE ROTATIONAL MOULDING OF LIQUID POLYMERS

It should be apparent from the preceding discussion that the rotational moulding of liquid polymers is a highly complex process. It is possible, however, to make the following recommendations[6] for minimising the number of problems likely to be encountered when processing such materials.

1. Choose a material with an initial viscosity in the range of 0.05→1.00 Pa.s and with a slowly increasing viscosity (3→4 Pa.s per minute up to 75 Pa.s and then rapidly increasing). For example, when moulding PVC plastisols, diisodecyl adipate plasticiser will produce a less rapidly increasing viscosity than dipropylene glycol dibenzoate.

2. Operate at the lowest possible rotation speed in order to minimise pool disturbance and bubble formation. This will also minimise the possibility of lifting the entire pool of a high viscosity material on one part of the mould wall during the first rotation.

3. Use moulds with generous corner radii (25 mm recommended).

4. Use a double or multiple shot process for thick walled mouldings.

5. Fillers such as glass microspheres may be added to the liquid in order to modify its flow behaviour. Ensure that the coating on the filler does not react with the liquid.

REFERENCES

[1] Schneider, Dr K., Fahnler, F. and Luttgens, B., 'Production of Large Capacity Hollow Vessels by Activated Anionic Polymerization of Lactam', Kunstoffe, V 58, Jan. 1968, p2-8.

[2] Schneider, Dr K., Keuerleker, R. and Fahnler, F., 'The Production of Rotationally Moulded Hollow Articles by the Activated Anionic Polymerization of Lactams', Kunstoffe, V58, 1968, p 1-5.

[3] Blockmann, A.F., 'Rotational Moulding - Recent Developments in Machine Construction and Moulding Procedure', Kunstoffe, V61, 1971, p 450 - 460.

[4] Blair, E.A., Monster Moulding Co, Canada. Patent 3, 976, 714 (Oct. 23, 1975).

[5] Blair, E.A., 'Rotomoulding Urethanes at Room Temperature', RPN Technical Notebook V22, July 1978.

[6] Harkin-Jones, E., 'Rotational Moulding of Reactive Plastics', PhD Thesis, The Queen's University of Belfast, May 1992.

[7] Throne, J.L., 'Reactive Rotational Moulding', Polymer Engineering and Science, V20,1980,P 899-P 919.

Index